Birkhäuser

Studies in Universal Logic

This series is devoted to the universal approach to logic and the development of a general theory of logics. It covers topics such as global set-ups for fundamental theorems of logic and frameworks for the study of logics, in particular logical matrices, Kripke structures, combination of logics, categorical logic, abstract proof theory, consequence operators, and algebraic logic. It includes also books with historical and philosophical discussions about the nature and scope of logic. Three types of books will appear in the series: graduate textbooks, research monographs, and volumes with contributed papers. All works are peer-reviewed to meet the highest standards of scientific literature.

Francisco Rodríguez-Consuegra

Russell, Gödel, Tarski

Selected Papers on the Philosophy of Logic and Mathematics

Birkhäuser

Francisco Rodríguez-Consuegra
Full Professor (Retired)
Valencia University
Valencia, Spain

ISSN 2297-0282 ISSN 2297-0290 (electronic)
Studies in Universal Logic
ISBN 978-3-032-08944-1 ISBN 978-3-032-08945-8 (eBook)
https://doi.org/10.1007/978-3-032-08945-8

Mathematics Subject Classification: 03-02, 03A05, 03-XX, 03AXX, 00A30, 03A10

This book is published under the imprint Birkhäuser, www.birkhauser-science.com by the registered company Springer Nature Switzerland AG
The registered company address is: Gewerbestrasse 11, 6330 Cham, Switzerland

Preface

This collection represents a compilation of some of the most important papers and essay reviews of my research pertaining to the contributions of Bertrand Russell, Kurt Gödel and Alfred Tarski to the philosophy of logic and mathematics. Also, I have assembled these papers to convey key elements of my research on the unpublished materials of the three greatest logicians of the 20th century. To my knowledge, I am the only scholar who visited and worked in depth in the Russell Archives (Hamilton, Canada), the Gödel *Nachlass* (Princeton, USA) and the Tarski Papers (Berkeley, USA).

This work is reflected in detail in the materials contained in this compilation. Resorting to archival unpublished materials has shown itself as a unique, indispensable resource to understand the actual thought and ideas of the three great logicians to which I devote this collection.

I have chosen them, not only for their common theme, but also because many are not easily accessible to modern scholars for reasons such as extinction of the hosting journal or publishers who are no longer active.

The papers trace my journey in my field in ways that connect to other scholars. Chapter 1 is focused on my study of Russell's theory of types. Its impact on the Russell Studies has been well documented, among them, its role in my initial controversy with Gregory Landini that ultimately led, in his revelation to me, to his developing ideas for a book on the same subject.

Chapter 2 is the full text a lecture to the Buffalo Logic Colloquium, November 1989. I was invited by John Corcoran while I was a Visiting Scholar in the Russell Archives, and the Philosophy Department, Hamilton University (Canada). Later, it was much expanded until reaching a whole book (*Relational Ontology and Analytic Philosophy: Bertrand Russell and Bradley's ghost*),[1] which remained unpublished

[1] Complete details of that book are here: https://www.amazon.com/dp/1709974710.

for a long time. The story of that book is well explained in the Preface and also in a paper of mine devoted to my personal relationship with John Corcoran.[2]

Although I started studying Russell's analytical method in my doctoral dissertation,[3] I was soon taken from Russell to Gödel then to Tarski. In a lecture about the importance of Russell's unpublished manuscripts, given in Hamilton University (Canada) in 1990, I tried to defend him from a typical accusation: Russell never understood Gödel's theorems (see Chaps. 3 and 4 of this volume). The late Gregory Moore was in the audience, and he advised me to visit the Gödel *Nachlass* in Princeton. That led me to devote several years to study those materials and later to publish papers and books about them. One of the results of that work was my thesis that Gödel was also a philosopher, who tried to use his celebrated technical achievements as a basis to support philosophical positions.

That line of research is developed in the three critical essay reviews contained in Chaps. 5–7. They constitute a full new interpretation of Gödel's philosophy, which was totally absent from the official *Collected Works*. That was admitted by the Editor in Chief, the late Prof. Feferman. There is a detailed study of my work on Gödel's philosophy, which includes my two books editing Gödel's unpublished philosophical manuscripts,[4] as well as those three essay-reviews: https://www.uv.es/rodriguf/Zal amea-Godel.pdf.

The path to Tarski reveals the serendipity of unexpected discoveries. One of mine was that Tarski's famous definition of logical consequence was anticipated by members of the Peano school. See Chap. 8.[5] The second was my belief that the celebrated Tarski Theorem, that arithmetical truth cannot be defined in arithmetic, was first formulated by Gödel, in a especially convoluted way (here Chap. 11). This took me to Berkeley from Spain to visit the Tarski Papers. There I discovered interesting materials of great philosophical interest. These studies involved a deep dive into Tarski's philosophical views on set theory, which resulted in writing a long a paper (included here as Chap. 12). It also led to the publication of several of Tarski's unpublished materials.[6]

The materials appear here as they were published, almost in the same chronological order. Yet I have taken advantage to correct a few errata, introduce a few minor

[2] "My personal memories of John Corcoran". In Jean-Ives Beziau and Tim Madigan (eds.), *Universal Logic, Ethics, and Truth: Essays in Honor of John Corcoran (1937–2021)*, Studies in Universal Logic, Birkhäuser, 2024, pp. 151–163.

[3] A significant part of that dissertation was reelaborated and published as *The Mathematical Philosophy of Bertrand Russell: Origins and Development*. Basel, Boston and Berlin, Birkhäuser, 1991 (reprinted 1993). To compose it, the heavy use of unpublished materials was, once again, indispensable.

[4] *Kurt Gödel, Ensayos inéditos*. Barcelona, Mondadori, 1994. *Kurt Gödel, Unpublished Philosophical Essays*. Basel, Boston and Berlin, Birkhäuser, 1995.

[5] Full details about that, in the context of the Peano School, can be found in the third chapter of the book, coauthored with E. Marchisotto and J. Smith, *The Legacy of Mario Pieri in Foundations and Philosophy of Mathematics*. New York and Basel, Birkhäuser, 2021. The contents of this book can be seen here: https://link.springer.com/book/10.1007/978-0-8176-4823-7.

[6] "Two unpublished contributions by Alfred Tarski." *History and Philosophy of Logic* 28 (2007), pp. 257–264.

improvements and some update in a few bibliographical references. As a sort of final reflection, I wrote a new chapter, formerly unpublished. It is an informal view of the way how some philosophers inspired by logic reach their most important achievements. At the same time, I take advantage to describe my personal relationship with some of them, often adding a little touch of humor, as a sort of contrast with the strictness of the former papers.

Now, already retired from research, I hope they may reach still a greater number of readers than they reached originally. These papers represent an important part of my work between 1985 and 2005. Together with the several books I devoted to these fascinating subjects, they constitute almost a whole research life.

I cannot finish without expressing my gratefulness to Elena Marchisotto, who encouraged and assisted me about this project; to Francine Abeles, Book Review Editor of *Modern Logic*, who proposed me to write critical studies of Gödel's *Collected Works*; to Jean-Yves Beziau, who approved the project for publication in his series Studies in Universal Logic, and to Frida Trotter, Birkhäuser Editor, who confirmed the approval and assisted me in a number of ways, to make a better book. Needless to say, my thanks are due to the journals and publishers where these papers appeared originally, for their kind permissions to reprint them.

Finally, my warmest thanks to my wife Elena, for his continuous and tireless support from many points of view.

Bédar, Spain
July 2025

Francisco Rodríguez-Consuegra
francisco.rodriguez@uv.es

Origins of the Materials

1. "Russell's theory of types, 1901–1910: its complex origins in the unpublished manuscripts". *History and Philosophy of Logic* 10 (1989): 131–164.
2. "The evolution of Russell about relations"
   Text of a lecture to the Buffalo Logic Colloquium, Nov. 1989
3. "Russell's unwritten reply to Gödel"
   Partial text of a lecture to the Russell Society, Hamilton, Canada, Jan. 1990
4. "Russell, Gödel and logicism". J. Czermak (ed.), *Philosophy of mathematics*. Proceedings of the 15th International Wittgenstein-Symposium. 16th to 23rd August 1992, Kirchberg am Wechsel (Austria). Vienna: Hölder-Pichler-Tempsky, 1993: 233–42. (Reprinted in A. Irvine (ed.), *Bertrand Russell: Critical Assessments*. London and New York: Routledge, 1998, 4 vols.; vol. 2: *Logic and mathematics*: 320–29.)
5. "Gödel's first works, 1929-1936: mathematics without philosophy?". *Modern Logic* 3 (1992): 58–74.
6. "Gödel's last works, 1938-1974: the emerging philosophy". *Modern Logic* 4 (1994): 318–26.
7. "Gödel's unpublished manuscripts, 1930–1970: the official edition". *Modern Logic* 6 (1996): 413–21.
8. "Definitions and logical consequence in the Peano School". *Theoria* 12/1 (1997): 125–37.
9. "Philosophy in Hao Wang's conversations with Gödel". *Modern Logic*. Finally appeared in v. 8, n. 3 & 4, May 2000–October 2001: 137–152.
10. "Propositional ontology and logical atomism". In Godehard Link (ed.), *One Hundred Years of Russell's Paradox. Papers from the 2001 Munich Russell Conference*. Berlin: de Gruyter, 2004, pp. 417–434.
11. "Is Tarski's Theorem really Tarski's"
    Much expanded text of a presentation to a Logic Conference, Madrid, 1993.
12. "Tarski's intuitive notion of set", in G. Sica (ed.), *Essays on the foundations of mathematics and logic*, Monza: Polimetrica (Advanced Studies in Mathematics and Logic, vol. 1), 2005, pp. 227–266.
13. "From Logic to Philosophy". Written for this volume.

Contents

Chapter 1
Russell's Theory of Types, 1901–1910: Its Complex Origins in the Unpublished Manuscripts

Abstract In this article I try to show the philosophical continuity of Russell's ideas from his paradox of classes to *Principia mathematica.* With this purpose, I display the main results (descriptions, substitutions and types) as moments of the same development, whose principal goal was (as in his *The principles*) to look for a set of primitive ideas and propositions giving an account of all mathematics in logical terms, but now avoiding paradoxes. The sole way to reconstruct this central period in Russell is to resort to unpublished manuscripts, which show the publications of these years to be the extremities of one same iceberg. Thus, the logical problems (doubts about propositions, matrices and functions) are parallel to the ontological (the searching for genuine logical subjects) and methodological ones (the status of eliminative reduction). Paradoxes were the cause that the kind of (constructive) definition already applied by Russell, as an inheritance of Moorean analysis, obtained a new trait: ontological elimination, which, beginning with a 'no classes theory' and the theory of descriptions, led to a new theory of judgment (already present in manuscripts from 1906), 'incomplete symbols', and logical constructions.

Keywords B. Russell · Paradox · Logical types · Logic · Mathematics · Philosophy

Mathematics Subject Classification Primary 0302 · Secondary 03A05

1.1 Introduction

The following is an attempt to reconcile changes and continuity in this important period of Russell's evolution, and also to provide a general view according to which the unpublished manuscripts throw light on the difficulties of the published theories of 1906, 1908 and 1910, which cannot otherwise be compared usefully. Thus, my use of the manuscripts is destined mainly to improve our understanding of the published works through the reconstruction of the *general* evolution of the historical Russell. The 'continuity' I mention is mainly a methodological one, in the sense that one of

F. Rodríguez-Consuegra, *Russell, Gödel, Tarski*, Studies in Universal Logic,
https://doi.org/10.1007/978-3-032-08945-8_1

my goals will be *to show* the actual method of Russell (as described in Sect. 1.2), which was maintained under different changes in the *content* of his doctrines. This provides a link with his former work, as I show in other of my papers (analyzing the manuscripts from 1898 to 1900 in 1988a; through the study of *Principles* and its background in 1988e and 1988d; and supplying an explanation of the unpublished origins of the theory of descriptions in 1988g).

I must point out, however, the extreme difficulty and the endless zigzags of the relevant manuscripts; in fact I think that one can prove almost anything from them. Further, I cannot supply here a *detailed* analysis of these manuscripts, which would transform this article into a whole book, but just enough analysis to maintain my general interpretation of the Russellian evolution from 1905 to 1910, including explanations of the philosophical 'changes of mind' in a general context coming from former periods.

The main line of Russell's evolution, *as may be seen only through the study of the manuscripts,* is as follows. The substitutional theory was developed technically departing from the theory of descriptions, but the mechanism of substitution required to make a difference between propositions and individuals (constituents), which led, already in a manuscript of 1906, to the abandonment of the old Moorean theory and to the admission of the propositions as *mere combinations,* that is to say, to the new theory of judgment as a 'multiple relation'. Therefore, the original theory had a consequence incompatible with one of its grounds, for the substitutional theory needed propositions to be genuine entities. On the other hand, there was another problem which added new difficulties to the already doubtful status of propositions: the semantic paradoxes, whose solution seemed to require a distinction between genuine propositions and mere statements (those containing apparent variables, namely, quantifiers). The role played by semantic paradoxes in the period 1905–1910 was the link between the problem of the universality of logic and the extension of the theory of types to the problem of the nature of propositions (to the point of claiming—in 1910—that they are also incomplete symbols). If we add the other difficulties in the substitutional theory, as exposed by Russell himself, both in published and unpublished writings (and pointed out here), we can understand why the manuscripts from this period were an experimental laboratory where Russell was continuously testing all the possible combinations with the three main elements he knew: propositions, matrices and propositional functions, trying to find a satisfactory theory.

Yet this theory was never found. The final pre-eminence was for propositional functions, for they can (among other things) eliminate classes; generate propositions; handle quantification in a way that makes possible (apparently) the avoidance of the distinction between proposition and statement; provide the material for the corresponding hierarchy of types (from which to derive the propositional hierarchy), and allow an easy axiom of reducibility (by means of the recourse to predicative functions). However, the version from 1908 had a terrible philosophical flaw (to my knowledge neither pointed out nor explained before): *it juggled with the status of propositions,* which are neither primitive nor derivative ideas. It is true that Russell was avoiding philosophical problems in 1908a, but he did not explain what particular problems were eluded. Now I suggest a new response: at this point he had a new theory of judgment which made impossible the admission of propositions, *but* he had not yet finished a general theory of incomplete symbols able to provide a new simplified ontology concerning propositions. This can explain why we find in 1908a traces of the distinction between proposition and statement (despite the new distinction between generalized and elementary propositions) and even of the old mechanism of matrices; and why the old Moorean ontology identifying terms and concepts was replaced by the subject-predicate pattern implicit in the notion of elementary proposition (see my 1988i). As explained below, only the final version of 1910a faced the problem adding explicitly the new theory of judgment to the fundamental hierarchy of propositional functions, although with the drastic (and never completely explained) solution of regarding propositions as incomplete symbols and, at the same time, admitting elementary propositions as a primitive idea.

In this way, I try to provide a whole view of the evolution of the actual Russell up to 1910 concerning an important topic. I am not concerned with the theory of types in itself (I think there is no such a thing), but only with the understanding of the historical theory of types *by Russell,* and in the role played by general philosophical and methodological problems in the development of this theory. (In my 1988g, which is a complement to the present article, I try to provide the same for the evolution of the theory of denotation from 1902 to 1905.) In the citations made by dating code, the prefix '*m*' to a piece by Russell indicates a text which is still unpublished [in 1989]: we have (for example) both 'm1906a and 1906a'.

1.2 The Role of Paradoxes and the First Theory of Types

In my 1987a, 532 *ff* I show that the general method of Russell from 1898, which I call *constructive definition* (independently from other uses of this expression), can be reduced to five traits. These definitions: (i) *analyze* the concept in 'simple' components; (ii) presuppose a *reduction* to these components; (iii) try to use *common language* as a guide; (iv) obtain *precise* meanings, and (v) cause a certain *loss of intuitiveness.* However, one might think that there is something lacking in the list: the *elimination* of the entity reduced to its components.[1] In fact, if we say that a concept has been *reduced* to certain others, it seems necessary to deprive it of its 'reality' because it has been possible to show it, by means of the definition, as only a set of elements which are ontologically simpler.

However, the Russellian line was different in *Principles* (1903a). He did not consider there that his definitions eliminate mathematical concepts because of: (i) the implicit Platonism that, resting on the being-existence distinction, grants ontological

[1] The ontological elimination proceeding from the theory of descriptions was a new *methodological* trait, which transformed the mere signs already present in the *Principles* into something systematic. The main difference was that in *Principles* the kind of constructive definition used by Russell did not eliminate ontologically the defined entities (as, for instance, numbers: see my 1987b). As for the origins of the theory of descriptions, see my 1988g.

reality to *all* that can be said or thought; (ii) the belief that the meaning of words are certain objects (their references), and (iii) the Moorean theory of judgment (inherited from Bradley: see my 1988h and 1988j), whose main trait was the non-distinction between the beings of simple concepts (the components) and of complex ones (the propositions).

Russell did not succeed in avoiding the problems implicit in the distinction between philosophical and mathematical definitions (see my 1987a, 538 *ff*) just because he was convinced it was possible to differentiate definitions as mere abbreviations from definitions as genuine reductions, and this tension remained without solution just because of the lack of an eliminative technique. This was the reason of his strange logicism: on the one hand he tries to *reduce* all to logical objects; on the other hand he grants the same kind of *being* for mathematical objects. This apparent incoherence disappeared only as a consequence of the discovery of paradoxes.

Russell arrived at his famous paradox trying to verify Cantor's theorem (according to which there is no greatest cardinal) with the class of all terms, which presumably must have the greatest possible number of members (1903a, §110). Reasoning on this possibility (1903a, §§348–349), he asked if, considering the class of all classes not belonging to themselves, this class belongs or not to itself, and discovered that each of the two possible answers led to the contrary one. That is to say, the class of all classes not belonging to themselves belongs to itself if and only if it does not belong to itself. As is well known, this was the paradox which he presented to Frege in 1902 (Frege 1976a, 131).

Cantor admitted that neither the class of all classes belonging to the class of all thinkable things, nor this last class, are true classes; so there are certain manifolds which are not actual unities: it is *impossible* the simultaneous existence (*Zusammensein*) of all its elements. For him, they are not genuine sets, but 'inconsistent systems' (see his letter to Dedekind in Cavaillès 1962a, 245–246). But Russell had many doubts over taking so drastic a decision, which would presuppose the abandonment of his theory of meaning and his ontology. This was the motive of his great resistance, as showed in some places of *Principles* where, although he admits the problem underlying the 'class of all entities' and similar difficult entities, he refuses to declare them as 'illegitimate', seeing this measure as 'undesirable' and 'desperate' (1903a, §344).

Such passages are the results of a battle against the admission of the paradox, which took place before publication of the last version of 1903a. As Coffa 1979a discovered, examining a discarded version of chapter 43 of this work (devoted to the infinite), the manuscript sent to the printer by Russell in May 1902 still contained an attack against the Cantorian demonstration that there is no greatest cardinal (in the 'hard' version of the diagonal method), departing from the belief that the class of all terms (the greatest possible class) does not verify Cantor's theorem. (For the earlier Russellian strong opposition against Cantor, see my 1988a.)

Probably this line of argumentation led Russell to handle the class of all classes not belonging to themselves.[2] Before it there were two possibilities: (i) to admit it can be proved that this class belongs to itself and the contrary; (ii) to declare this entity as non-existent, thus rejecting the attack against Cantor's proof. Obviously Russell chose the second one, and at this point the very reason of all the later ontological changes appeared. Thus, between June and November of 1902 he sent to the printer new materials for *Principles* relative to Chap. 10 (on the paradox), 43 (mentioned above) and the appendices (one of them on the first theory of types) (see Clark 1975a, 99–101). At last, in May 1903 he sent to Frege the first attempt of a 'no classes theory',[3] the result of the new changes. The main philosophical consequence of all that was the bankruptcy of the old theory identifying meaning and reference (see my 1988g).

All mentioned materials were written under the conviction that something supposed to be existent must be deprived of this existence. So, when regarding the possible ambiguity of the membership relation, he concludes that the class of all classes (w) leads to contradiction, and 'we must suppose, either that there is no such entity as w, or that there is no propositional function satisfied by its terms and by no others' (1903a, §103). The same happens with the general notion of 'class as one': 'if a collection of terms can only be defined by a variable propositional function [that in which the function itself can serve as argument], then, though a class as many may be admitted, a class as one must be denied'. This, again, leads to the notion of 'meaningless' (§104):

> there are propositional functions $\varphi(u)$ in which u may be the class as many, which are meaningless if, for u, we substitute one of the terms of the class... In this view, a class as many may be a logical subject, but in propositions of a different kind from those in which its terms are subjects.

This last distinction led to the theory of types and the no classes theory; but the main difficulty in accepting such views was that they involve an attack both against the identification between meaning and reference and against the true–false distinction, leading finally to a theory of judgment unable of admitting 'negative facts'. The last paragraph of chapter 43 seems to announce that the only solution consists in accepting both the non-existence of the greatest cardinal and the theory of types, denying 'that there are true propositions concerning all objects or all propositions', which seems to him 'plainly false, since all propositions are at any rate true or false' (§349).

---

[2] I have no place to examine in detail the historical problem of the precise origins of Russell's paradox, but I must mention that, before Coffa's discovery was made, I. Grattan-Guinness proposed another likely explanation based on the correspondence with Hardy (see his 1978a and his judgment of Coffa's explanation in 1981a). The following two articles are also interesting: Moore/Garciadiego 1981a and Garciadiego 1987a.

[3] When I speak about the *first* no-class theory I mean mainly the first no-class *theory*, namely, the first serious attempt to put at work a general idea to the point of building logic and mathematics. But I do not mean that there was no former suggestions in the same line, as I point out in Sect. 1.2 and footnote 5 below.

The first theory of types, as stated in 1903a (appendix B, §§497–500), contained already some of the elements of the later theory, in particular the notion of *range of significance* (§497):

> Every propositional function $\varphi(x)$—so it is contended—has, in addition to its range of truth, a range of significance, i.e. a range within which x must lie if $\varphi(x)$ is to be a proposition at all, whether true or false. This is the first point in the theory of types; the second point is that ranges of significance form *types*, i.e. if x belongs to the range of significance of $\varphi(x)$, then there is a class of objects, the *type* of x, all of which must also belong to the range of significance of $\varphi(x)$, however φ may be varied; and the range of significance is always either a single type or a sum of several whole types.

The resulting hierarchy is as follows. First type, *terms* or *individuals*, which are not ranges, being the lowest kind of objects; second type, classes (or ranges) of individuals; third type, classes of classes of individuals, and so on. In general, the type $m + 1$ is formed of all classes of objects of the type m. When relations (in extension) are introduced, the theory becomes complicated, with ranges of relations, relations of relations, relations of couples, relations of individuals to couples, and so on: 'in this way we get not merely a single progression, but a whole infinite series of progressions' (1903a, §497).

With all that certain paradoxes are solved, as for instance those concerning classes, Cantor's paradox, and similar others (see Copi 1971a, 24); but Russell was able to construct a different kind of paradox concerning *propositions*. Let m be a class of propositions, w the whole class of propositions of the form 'every m is true' having the property of not being members of m, and p the proposition 'every m is true'. Then: 'If p is a w, it must possess the defining property of w; but this property demands that p should not be a w. On the other hand, if p be not a w, when p does possess the defining property of w, and therefore is a w' (1903a, §500).[4]

Russell claimed that this kind of paradox was 'unavoidable' and he was aware that they required 'further subtleties for its solution', adding: 'but what these are, I am at a loss to imagine' (1903a, §500). Of course, the problems concerned were the 'semantic paradoxes' (as they were called later), whose main trait is the 'vicious circle' we commit by the self-application of general propositions (see Sect. 1.4 below). Russell was also aware that the kernel of the problem was the need of regarding propositions as a new type from which to form a new hierarchy, but claimed: 'there are certain difficulties in the way of such a view, which render it doubtful whether propositions can be treated like other objects' (1903a, §497). As we shall see, he was right in reducing the question to an ontological problem, which started considering the possibility of dispensing with classes.

The first exposition of a no classes theory, in a letter to Frege of 24 May 1903 (Frege 1976a, 158–160), shows clearly that Russell had decided to make great ontological

[4] In this presentation I follow narrowly the historical Russell, but a clearer version of this paradox (for which I am indebted to G. Landini) is as follows: For every class m of propositions there is a unique proposition 'every m is true'. Consider now the whole class w of propositions of the form 'every m is true', and having the property of not being a member of their respective m's. Then there is a unique proposition 'every w is true', and this proposition will be in w iff it is not in w.

sacrifices[5] and, at the same lime, a constant both in substitutional theory and in theory of types: the discovery that classes are completely superfluous, which might allow the construction of an arithmetic with no classes avoiding so the paradoxes. It consists in dispensing with the habitual notation for classes in terms of the notion of 'such that', admitting instead of it expressions with names of functions, as for example '$\varphi \supset \psi$'. This theory was finally abandoned because it was unable to avoid paradoxes (as Frege said in his response, protesting against Russell's use of names of functions as isolated entities, so forgetting their unsaturated character).

Departing from the new idea of dispensing with classes, which underlies all subsequent progress, the next main steps were: (i) theory of descriptions; (ii) substitutional theory (with the new theory of judgment); and (iii) theory of types. As I have studied the first step elsewhere (see my 1988g, which shows the first seeds of substitutional notation as well as the need of a new theory of denotation in order to solve paradoxes), I shall try in the following of explaining the rest, resorting to both published and unpublished writings, and using as a general guide the methodological point of view: the constructive definition (see above) and the ontological assimilation of the eliminative element.

1.3 From Descriptions to Substitutional Theory

In 1944a, 13–14, Russell explains how, desperate by the paradox of classes, he saw in the problem of descriptions some hope, and how the solution to it suggested 'a new technique' by means of the discovery that descriptions were meaningless when separate from meaningful statements which contained them: 'It soon appeared that class-symbols could be treated like descriptions, i.e. as non-significant parts of significant sentences. This made it possible to see, in a general way, how a solution of the contradictions might be possible.' During a long time this passage has been a sort of mystery because the relationship between incomplete symbols and theory of types ('the solution' in *Principia*) was not clear enough (see for instance, Quine 1966a, 661). With the posthumous publication of the manuscript 1906b, containing the first complete version of substitutional theory in Russell 1973a, and the availability of the rest of relevant manuscripts (in The Bertrand Russell Archives), the mystery can be completely clarified, developing the line opened by Grattan-Guinness 1974a, who pointed out the substitutional theory as the link between descriptions and types.

It can be said that theory of descriptions allowed the practical application of the idea according to which it is possible to dispense with classes: the no classes theory (see above). However, it did not suppose the substitution of propositional

[5] The idea of dispensing with classes as entities was related with other similar possibilities, as is shown in the correspondence with Frege. For example, the belief that classes cannot be admitted like proper names (letter of 10 July 1902: Frege 1976a, 137), or the first formulation of the theory of types, which, distinguishing a relation between relations and a relation between objects as belonging to different types, comes to call 'nonsense' any attempt to mix both levels (letter of 8 August 1902: p. 145).

functions for classes, which implied the same danger, but the need to imagine a device dispensing with both: substitutional theory was this recourse, at least as a first purpose.[6] In fact, it had been introduced already as a new notation (see my 1988 g), but it was abandoned because of the problems caused by the distinction between meaning and denotation (among others). As these problems had disappeared with theory of descriptions, the old idea was again possible. Theory of types, which supposes a more conservative notation, continued the spirit of no classes theory and (in some points) of substitutional theory itself (see Landini 1987a). Thus, the correct general sequence consists in seeing both substitutional theory and theory of types as *two forms* of the no classes theory,[7] and, in particular, substitutional theory as a generalization of the paraphrastical methods from 'On denoting'.

The preserved correspondence otters the first signs of this generalization in autumn of 1905, after several discarded attempts. Thus Russell said the following to Moore on 25 October 1905, concerning the possibility of reducing the problem of denotation to the problem of the variable:

> Carrying on a similar method in the analysis of propositions, I am now trying to work out a theory according to which there are no classes. In this view, propositions *verbally* about the class *mankind* are really about all men or some men. This would (with a similar doctrine for properties, except straightforward predicates) solve the various contradictions. It seems broadly possible to maintain that there is nothing complex but propositions.[8]

The central idea was certainly the same: in the same way that descriptions without reference showed the need of dispensing with all denoting expressions (at least from 1905a), so classes leading to paradoxes (presumably as well as a lack of reference)

---

[6] Russell was completely aware of this from 1903a, where he saw how the paradox can be formulated equally with such functions. However, the things exactly rejected in 1903a, 88, are not propositional functions as logical subjects, but only the functional parts of propositional functions. Besides, in 1903a: (i) classes are derived from propositional functions (p. 88); (ii) we can read that 'propositional functions must be accepted as ultimate data' (p. 88); and (iii) propositional functions are *primitive ideas,* which must be regarded as the genuine and indefinable residue of the process of philosophical analysis, that is to say: they are fundamental logical concepts, irreducible and simple, with which the mind has 'that kind of acquaintance... which it has with redness or the state of a pineapple' (preface, xv: see the—to my knowledge—only study on the indefinables of 1903a in my 1988c). Anyway, though he wrote that the existence of classes and relations can be objected to in the same way as 'the existence of propositional functions as separable entities distinct from all their values' (1906b, 154), and though in this article Russell interprets the expression 'propositional function' as an abbreviation of some or all their values, he was mainly looking for a way to avoid the speaking about quantification over values of the functional part of the function (φ), eluding so 'contradictions'. The result was a method to translate functional notations into matricial ones (1906b, 188; but the article was not published until 1973a); and it finished by pointing out the danger implicit in the assumption of propositions as entities, whereas in the exposition of the theory actually published (1906c), devoted to solve especially semantic paradoxes, there is no mention of the possibility of replacing propositional functions by matrices.

[7] Bunn (1980a, 241), exceptionally, formulates the thesis in this way.

[8] The same in the letter of 2 November 1905 to Jourdain, asserting to be capable of solving the contradictions: 'I do so by wholly denying that there are such things as classes and relations, which I treat in the same way as I treated denoting phrases in the current *Mind*' (in Grattan-Guinness 1977a, 56).

suggested the need of dispensing with all classes: 'I think it is demonstrable that inconsistent classes are not classes or anything else, but are merely mistakes. I now extend this to *all* classes' (letter to Jourdain of 1 January 1906, in Grattan-Guinness 1977a, 68).

The first attempt of a substitutional theory may be found in 1906a (from November 1905), together with brief explanations of two other theories: *zigzag* and *limitation of size,* which were abandoned because of (among others) excessive complications, a certain ad hoc character, and lack of clearness (see 1906a, 146*ff*). The third theory was the substitutional one, although still under the name of 'no classes theory'. It is introduced as one among the others (1906a, 156), but in an added note (from February 1906) Russell wrote that this theory must be considered as 'the complete solution'.[9]

The exposition offers only the essentials: to show the paraphrastical analysis of classes in a way that they cannot be seen as unities. With this aim, Russell omits the usual notation of propositional functions, which 'inevitably suggests the existence of something denoted by "φ"'. Instead of it Russell introduces the notation '*p(x/a)*', where *p* (the *prototype*) is any proposition and *a* (the *origin* or *initial subject*) is a constituent of *p,* which can be read: 'what *p* becomes when *x* is substituted for *a* in the place or places where *a* occurs in *p*'. Immediately this recourse is applied to classes; thus, instead of 'the class *u* is a class which has only one member', we must say 'there is an entity *b* such that *p(x/a)* is true when, and only when, *x* is identical with *b*. In this way, 'the values of *x* for which *p(x/a)* is true replace the class *u,* but we do not assume that these values collectively form a single entity which is the class composed of them' (1906a, 155).

In 1906c the parallelism with descriptions results still more explicit, when Russell starts from the notation *p/a;b,* defining it as 'the *q* which satisfies *p*(*b/a*)!*q*': 'Strictly speaking, as in all classes of phrases beginning with *the,* we do not define *p/a;b* itself, but we define any statement in which it occurs': therefore, the statement '*p/ a;b* has the property φ' is to be defined as 'there is a *q* such that *p*(*b/a*)!*r* is true when and only when *r* is identical with *q,* and *q* has the property φ' (*1906c,* 200–201). Departing from that, the constructive definition, now incorporating the eliminative element, might be extended to all foundations of mathematics.

However, the power of this application of the new method can be seen in its clearest way in the symbolical formulation of *m1905a* (dated 22 December 1905). This manuscript was a first attempt to develop a system of axioms (Pp) and theorems to base mathematics in the new theory. It starts from a clearer idea of *substitution* (between constants) as a process different from *determination* (a constant in the place of a variable), a distinction missing in former attempts. The basic definitions of descriptions and existence of the individual described (still with the old notation with the sign of admiration) are then applied to the elimination of *p/a,* a non-explained notation destined to transform itself into the sign of the later called *matrix,* a substitute

[9] The letter from Couturat of 12 December 1905 shows that Russell felt certain that he had solved the contradiction formerly: 'Je suis ravi d'apprendre que vous trouvé la solution de la contradiction' (Russell Archives).

for classes (m1905a, *12.112.114, fol. 4):

$$\varphi!\iota`\overline{p/a}. = \therefore(\exists b)\therefore p(y/a)!q. \supset_{q,y} : q. \equiv .y = b \therefore \varphi!b \text{ Df},$$

$$\text{Ex}!\iota`\overline{p/a}. = \therefore(\exists b)\therefore p(y/b)!q. \supset_{u,y} : q. \equiv .y = b \text{ Df}.$$

The manuscript continues with more Pp and theorems, finishing with dyadic substitutions, destined to replace relations. However, this axiomatic structure had no continuation, perhaps because it allowed a certain mixture of matrices and propositional functions (see Grattan-Guinness 1974a, 390–391). We shall see below that propositional functions appear as well even in the version of substitutional theory devoted to semantic paradoxes.

1.4 Substitutional Theory and the Ontology of Matrices

The notion of *matrix* is just the one used in 1906b (read in April) to generalize the concept of 'incomplete symbol' (expression not yet mentioned), still implicit in 1905a, to the rest of entities except to propositions and individuals. The 'fundamental logical principle' which, according to Russell, rules this generalization is only a version of Fregean context principle: 'in any sentence, a single word, or a single component phrase, may often be quite devoid of meaning when separated from its context' (1906b, 165). When, in these cases, some meaning is erroneously assumed, we have 'false abstractions' or 'linguistic fictions' (pp. 166–167), with which Russell regains the respective terminologies from Bradley and Meinong. Russell mentions as such pseudo-entities classes, relations, numbers and 'almost all the things that mathematics deals with' (p. 166).

There is, however, a difference: whereas 'the present King of France is bald' is false, 'the number one is bald' is meaningless (it does not express a proposition). The reason Russell offers is that 'there are really no such things as classes'; so, the statements containing them are meaningful only when analyzed into other statements 'about all or some of the members of the class' (p. 166). But this reason is not very convincing and it seems that the real problem is rather ontological than syntactical: Russell is implicitly referring to the impossibility of applying to an entity properties of different kind, as is demonstrated by the fact that he mentions, as examples of unacceptable propositions, 'the class of human beings is an human being' and its negation. In saying that I am not forgetting that substitutional theory can dispense with classes, but emphasizing the difference between classes and numbers, which must previously be shown to be classes by means of an *independent* logicist definition (see my 1987b). Anyway, in other places Russell calls the method of substitution 'a technical device' (1906c, 200), coinciding so with the fact according to which, in finishing the exposition of the theory, he admits that all the results obtained by means of this method would hold even with existing classes and relations: 'we do not deny that there are such entities, we merely abstain from affirming that there are' (1906b, 188).

As explained above, matrices are to replace classes (including as well all entities that can be defined from them, like numbers), propositional functions and relations. In any substitutional pattern *p/a;x, p/a* is a *matrix* containing a symbol of an individual *a* and that of the proposition *p* where the former must replace to *x* (although we are told it is not essential to think of *p* as a proposition, but only as the name of 'a genuine entity' (1906b, 168). Therefore a matrix is not an entity, but an *operation* that, like other mathematical symbols, acquires a meaning only as a part of certain propositions. Thus, neither the 'shadowy symbol' (as Russell wrote) *p/a* (a class), nor *p/(a,b)* (a dyadic relation), are entities, and we arrive so to an assertion with the same sense that the corresponding one used in the case of descriptions: 'the matrix *p/a* is a symbol for the phrase "the result of replacing *a* in *p* by", which is incomplete and meaningless; in order to acquire a meaning, we must add the name of the entity which is to replace *a*' (p. 170).[10]

Starting from this point, the theory is developed easily. The rejection of propositional functions allows to avoid the problem (unsolved in *Principles*) of quantifying over them (see, however, below). Now the function is nothing when separated from its values, but matrices allow to express the same idea avoiding the danger of paradoxes (1906b, 171–172). It is necessary, therefore, to regard substitutional theory as containing a theory of types (Landini 1987a, 174–175). In fact, instead of the variable function now we have two variable entities *p* and *a,* with which we can define membership avoiding paradoxes: 'To say that *x* is a member of the class *a* is now to say that for some values of *p* and *a, a* is the matrix *p/a* and *p/a;x* is true'; therefore '"*x* is an *x*" (or its negation) becomes meaningless, because "*x* is an *a*" requires that *a* should be of the form *p/a,* and thus not an entity at all' (1906b, 172).

We have the same with functions of two arguments, that is to say, with relations in extension (pp. 173–174). Russell introduces types explicitly according to the number of constants to replace; thus *p/a* belongs to the first type, *p/(a,b)* to the second one, etc. Yet he admits that it is not possible to define the meaning of 'type' because it would require a general statement about types and, as in the future theory of types, 'statements which are significant for one type are in general not significant for another'. Therefore statements in terms of matrices, which are only symbolic abbreviations, will be significant only if they fulfil the requirement of homogeneity of types, as for instance does *not* occur in $p/a = q/(b,c)$ (or its negation) (pp. 176–177).

The definition of cardinals starts from the old idea of regarding them as classes of classes, which is maintained (see above). 0 is defined as the matrix *q/(p,a)*, when *q* means '*p/a;x* is false for all values of x', that is: {(*x*). – (*p/a*);*x*)/(*p/a*)}. Similarly,

---

[10] The incompleteness makes matrices very like propositional functions, but Russell emphasizes the greater generality of matrices by comparing the two following statements, calling 'prescription' to the first one, and statement 'truly general' to the second one. The transcription was (1906b, 188):

$$(1)\ `x = y. \equiv: \varphi x. \supset_{\varphi} .\varphi y`; \quad (2)\ `x = y. \equiv: p/ax. \supset_{p.u} .p/ay'.$$

to say '*p/a* is a unit class' means 'it is not false for all values of *c* that, for all values of x, *p/a;x* is equivalent to "*x* is identical with *c*"' (175–176). In general, a class of classes is a dyadic relation defined by a function of *p* and *a,* where these appear only as *p/a.* That is: 'The matrix *q/(p, a)* is called a *class of classes* if, for all values of r, c, r', c', provided $r/c = r'/c'$, then $q/(p, a)$; (r, c) is equivalent to $q/(p, a);(r, c')$' (p. 176).[11]

Such definitions of cardinals, according to the hierarchy of types, are admissible only for one type: that of classes of entities; so they must be repeated for each type, to fulfil the expressed conditions of significance. When doing so, we avoid the problems involved in speaking of 'the total number of objects, entities, matrices', etc., trespassing so the barriers between types. Thus, it continues being true that $2^{\alpha} > \alpha$, where α is the total numbers of entities, but now there is no class *of entities* with 2^{α} terms. This solves the greatest cardinal problem (Cantor's paradox) as well as the paradox of classes: as a class is no more an entity, classes cannot be subclasses of entities. Now any two entities *p* and *a* define a class *p/a* (a matrix), but 'a pair of entities is not one entity', so the number of such pairs can be, with no contradiction, greater than the number of entities (1906b, 178–180).[12] Similar considerations are applied to solve the ordinals paradox: the ordinal number of a series is always of higher type than its members (pp. 180 *ff*).

However, the substitutional method (based on the distinction between proposition and constituent) presupposes no distinction between propositional substitution and individual substitution, as for instance when introducing *q/(p,a)* as a class of classes and *q/(a,b)* as a dyadic relation. This seems to consider propositions as individuals, but then the meaning of *p/a* and *a/b* is thoroughly doubtful. This is why Russell claims that his theory requires only to regard *p, q,* etc., as 'entities', but not necessarily propositions (1906b, 175 note). In this way we can see the substitutional theory as the last attempt to go back to the old Moorean ontology of 'terms', simple or complex, and we can describe it as a return 'to the pristine simplicity of the unrestricted variable and the univocity of being' (Cocchiarella 1980a, 81), at least in principle (however, see below).

Besides, there was another problem: the substitutional theory solves all paradoxes concerning classes and relations, 'but in order to solve the *Epimenides* we seem to need a similar doctrine as regards propositions' *(1906c,* 204). Thus, the final destiny of substitutional theory would be complicated by the same source that produced the 'ramifications' of the theory of types: the 'semantic' paradoxes.[13]

Russell himself pointed out very soon some problems of substitutional theory: (i) it seems obvious that there are classes; (ii) if there are no classes, a great part of Cantorian set theory falls down (with which a great part of accepted mathematics

---

[11] For technical details, see Grattan-Guinness 1974a and Landini 1987a.

[12] See below on a conflict between substitutional theory and Cantor's theorem claimed in Landini 1987a.

[13] Of course, I do not mean that the two 'ramifications' are the same one, although I think it is interesting to point out the similarities: (i) some hierarchy: (ii) certain limitations of significance; (iii) the use of propositional functions; and (iv) an axiom of reducibility. See below.

cannot be constructed); and (iii) its practical development is very complicated (1906a, 154). But in the canonical exposition of the theory he pointed out only a serious danger: the possibility of any other paradox as a result 'from the assumption that *propositions* are entities', because 'it is very hard to believe that there are no such things as propositions' (1906b, 188). The unpublished manuscripts show that, to give an account of semantic paradoxes, it is necessary to introduce great modifications in the original theory, which led finally to its abandonment; that is to say, to its presentation under the form of a theory of types eluding the more problematical aspects (1908a), and arriving finally to the point of dispensing officially even with propositions (1910a).[14]

Before discussing the reasons of this change, it is necessary, however, to expose the modifications introduced by Russell in published writings. In 1906c Russell begins by admitting that the key to solving the paradoxes is the vicious circle principle of Poincaré, according to which anything implying *every, any* or *some* members of a class cannot be, again, a member of this class.[15] As Russell says, 'Whatever involves an apparent variable must not be among the possible values of that variable' (1906c, 198); which, in the relevant case, will be; 'Whatever involves *all* must not be one of the *all* which it involves' (p. 204). Here Russell resorts to his distinction between *statement* and *proposition,* which was finally the main device to solve the problem. As any statement about *all* propositions involves a proposition as apparent variable, Russell modifies the habitual sense of 'proposition', redefining it as 'what is affirmed by a statement containing no apparent variable' (pp. 204, 207). In this way the vicious circle is avoided by distinguishing between the statements which affirm propositions and those which do not affirm them.

To achieve it. Russell resorts to his (implicit) notion of 'incomplete symbol' by means the concept of 'range of significance' of a function, which allows to make a difference between *individuals, classes. classes of classes, dual relations,* and so on. The problem caused by propositions stating such ranges is that they do not succeed in restricting correctly the variable. When we try to do it, for example with 'φx is only significant when x is a class', there appears the proposition 'φx is not significant when x is not a class', which eludes the restriction (1906c, 205–206). To avoid it, Russell sees only one way: to use variables whose 'internal structure' can state its logical type when different from that of individuals. Thus, we must limit possibilities of single letters to stand for individuals, that is to say, to admit only individuals as

[14] We must not forget the role of Whitehead in this evolution: mainly his criticisms against the excessive 'formalism' of substitutional theory, which reduces mathematics, according to him, to a mere typographic device (in contradiction with Russell 1903a). See the quotations of the relevant correspondence by Lackey in Russell *1973a,* 131–132.

[15] I. Grattan-Guinness has pointed out to me the lack of an explanation of the reason by which Russell chose just the vicious-circle principle from Poincare, which he handled since 1904. I do not know the reason, but anyway it is not strange that Russell, as in many other times, returned to former positions previously discarded. In this case the principle permitted the avoidance of self-application problems similar to those which led to reject the semantic dualism from Frege in 1905 (see my 1988g).

entities. Thus, the variable remains with no restriction because there is nothing not being an individual (p. 206):

> Hence to reconcile the unrestricted range of the variable with the vicious-circle principle, which might seem impossible at first sight, we have to construct a theory in which every expression which contains an apparent variable (i.e. which contains such words as *all, any, some, the*) is shown to be a mere *façon de parler*... And such expressions include all descriptive phrases (the so-and-so), all classes, all relations in extension, and all *general* propositions, i.e. all propositions of the *form* 'φx is true for all (or some) values of x'.

Thus, the program to generalize the concept of incomplete symbol in its end involves also propositions containing apparent variables.

The theory is applied immediately to semantic paradoxes. The *Epimenides* is solved interpreting 'I am lying' as 'there is a proposition p which I am affirming and which is false', which is a statement with an apparent variable but does not state a definite proposition. Therefore the statement is false and the liar is lying. The possible way out of saying 'I am making a statement which is false' fails, because we cannot make statements *in general,* but only statements containing a specified number of apparent variables. And if we want to point out this total number, we must do it in terms of a propositional function with the same number of variables; thus, we exceed always by one that number (φ is also a variable) and the statement cannot be applied to itself (1906c, 207–208). We obtain the same result with the paradox of the excluded middle, and similar considerations are applied to other paradoxes.

As I pointed out above, there is a great similarity between this theory and the later one of logical types. In both theories a hierarchy is stated, as well as certain limitations of significance according to it (although the idea of ranges of significance, already present in *Principles,* was developed rather in 1908a than in 1906a). The similarity increases when Russell verifies that, with the stated limitations, it is not possible to quantify over 'all propositions' without specifying one type in particular, which makes mathematical induction impossible and with it a great part of mathematics. The reason is simple: mathematical induction requires that 'any property' belonging to 0 and also to the successor of any number, belongs to all numbers. Therefore we must restrict the expression 'any property' to properties which can be stated without apparent variables, and it can be obtained by the following axiom (named later 'of reducibility'): 'every statement containing x and an apparent variable is *equivalent,* for all values of $x,$ to some statement φx containing no apparent variable' (1906c, 212). With this axiom we conclude that any matrix which has a statement as prototype is equivalent to another one whose prototype is a proposition; in this way we quantify over genuine logical subjects, that is to say individuals and propositions (see Landini 1987a, 183). We shall see below, however, that there was an insurmountable difficulty in this last distinction.

Yet there is another important similarity between 1906c and 1908a: in both we find propositional functions used. This is very surprising because matrices were supposed to replace propositional functions, as in 1906b. But to my knowledge there is no place in 1906c where Russell says that matrix notation *must* replace the functional one, not even that propositional functions must be avoided in any sense. And this is very important, for *1906c was published at* this time (in September 1906), whereas 1906b was only read (on 24 April and 10 May 1906), and has been published only posthumously: so we infer that 1906c must be regarded as a more definitive version. Here Russell solves standard paradoxes concerning classes by means of matrices, and tries to *apply the same method to propositions,* by introducing the distinction between genuine propositions and mere statements (1906c, 204). But he needed propositional functions to build *general* propositions (statements) by means of quantification over the values of the variable in the function. In fact the whole article is full of functional uses, and we can find even an example of quantification over them ('There is a propositional function... such that...: p. 208), in which they seem to appear as logical subjects, and not only as a convenient notation. In return, matrices are used in both articles to replace classes (1906b, 170; 1906c, 200–201). Thus, although it is true that propositional functions determine classes, these last supposed entities are to be eliminated by the 'simple' substitutional theory.

Summing up: from 1906b to 1906c[16] propositional functions seem to regain some of their old status in 1903a (and m1904a) as indefinable and ultimate entities (or primitive ideas of the system). As we shall see below, we can find the same in 1908a (with some doubts which we shall discuss) and 1910a.

1.5 The Abandonment of Substitutional Theory and the Manuscripts

In the literature we find several hypotheses concerning this change of mind. I shall say something concerning these attempts before presenting my study of the unpublished manuscripts.

Grattan-Guinness (1974a, 399) proposes as a main cause the appearance of the new 'multiple' theory of judgment, which destroys propositions as abstract objects by considering them as mere complexes, so eliminating the base of the former theory. Of course, both theories are obviously incompatible but, as we shall see below, there are as well textual evidences for regarding the new theory of judgment and truth rather as a *consequence* of the very problems related to the substitutional theory, than as a *cause* of the abandonment of this theory.

Hylton (1980a, 22 *ff*) protests against this explanation, but speculating upon the chronology of the abandonment (and the rejection of propositions) regarding only dates of publications. He offers positive reasons as well: (i) the theory permits also semantic paradoxes motived by the need of using the notion of truth, and leads finally to contradictions related with the notion of membership (1980a, 24); (ii) the solution of 1906c violates the intended universality of logic, which led to the substitutional theory itself (1980a, 25–26). But these two reasons are doubtful, in spite that both are used by Hylton to explain an *actual* fact: the final pre-eminence of propositional functions. Concerning (i), Hylton says that the new possible contradictions explains

[16] In 1906, published before 1906c, we find also propositional functions as a primitive idea (still in terms of the 'denoting complex').

'presumably' that Russell did not publish his 1906b until achieving some solution to semantic paradoxes. But it is obvious that when he found this solution in 1906c, the main article remained unpublished. Moreover, the solution from 1906c contained very different ideas (as we saw above), and some of them were the result of unpublished discussions (as we shall see below), which contain some signs of the future destruction of propositions as entities (as made in m1906b).

As for (ii), Hylton claims that when Russell states the restrictions motived by the orders of propositions, statements which should state such restrictions of significance should violate themselves, unless the unrestricted variable is abandoned, with which the main advantage of substitutional theory is lost. But, as we saw above, the theory from 1906c succeeds in solving semantic paradoxes without any restriction of the variable (Landini 1987a, 190 has seen this also). It would be enough with the distinction between statement and proposition to avoid that general propositions play the role of individuals or 'logical subjects'. In being rejected as mere statements ambiguously denoting a class of propositions they do not violate any restriction: they are mere *façon de parler.* However, the generalization of incomplete symbols (still implicit) was so strong that not even the abandonment of the substitutional theory allowed finally to preserve the ontology of propositions.

Landini 1987a, 185 *ff*, has introduced a new reason of the (according to him apparent) abandonment: the change was motived by a certain conflict between the substitutional theory and Cantor's theorem. I shall not enter in the details, but the essentials of his argument seems to be these two steps: (i) Russell had a lapsus in asserting that his theory eliminated the greatest cardinal paradox (see Sect. 1.4 above), which cannot be true because at the same time he maintains that the number of entities is infinite, and so we can prove that the number of pairs of entities is identical to the number of entities; (ii) departing from that and by means of matrices we can arrive at a paradox, which allows Landini to infer that, with the unrestricted variable, the matrix determining a class cannot involve a pair of individuals without paradox: 'Thus, either the apparent variables of quantification must be confined within a given range, or the statement containing apparent variables are not to be regarded as designating single logical subjects' (p. 186). Since Russell starts from the unrestricted variable, Landini concludes that, after all, substitutional theory must admit that, in a matrix of the form *p/a* the prototype can be a statement, that is to say, something not being a genuine individual. Therefore the axiom of reducibility, which tries to make equivalent matrices containing statements and matrices containing propositions, is no more effective. It led, according to Landini, to the theory of types of 1908a, where Russell stated a hierarchy of propositions which may contain apparent variables and also may be regarded as logical subjects of a given type, while individuals lack all complexity. Thus, they cannot be replaced by propositions, and the conflict with Cantor's theorem does not appear.

Landini's theory is likely and interesting, but it has the same defect as Hylton's one: it ignores some relevant unpublished manuscripts.[17] Anyway, as we shall see below, I cannot agree with Landini's interpretation of the role played by propositional functions in 1908a, and I cannot accept the weight that he concedes to the substitutional theory in this same paper and even in 1910a

At this point, I can try to propose my alternative explanation. In the following, as well as in the next section, I shall introduce new data from unpublished material with the aim to understand better the evolution from substitutional theory to the published theory of types, which will be considered in Sect. 1.7.

To my knowledge the relevant manuscripts are mainly two: m1906a, from April–May (so it is between 1906b and 1906c); and m1906b, from September and corrections of 1907 (between 1906c and theory of types of 1908a). As I pointed out in Sect. 1.1, they are working materials without form and continuity; so there are many difficulties and some risk in extracting clear ideas from them. Russell usually considers accurately the difficulties of the several theories he handles, but without thinking of any reader. Moreover, as the main line of thought (which I shall try to extract) is not explicitly formulated, it is necessary to make the reconstruction starting from the claimed purposes of other writings (former and later) of the same period.

In the case of 'On substitution' (m1906a) the principal purpose of Russell was the application of his new substitutional theory to semantic paradoxes, in a similar way than 1906c. However in m1906a he deals with many problems not appearing in the publication, which was devoted to expose the obtained results in combination with the vicious-circle principle. A close examination of these problems will allow us a better understanding of the difficulties of the theory and the beginning of the evolution towards final types

The manuscript starts with the 'form' of the proposition, which appears many times: 'Every proposition can be divided into a subject and a predicate, and in general this division can be effected in many ways' (m1906a, fol. 1). Such a categorical assertion, contrary both to the Moorean theory of judgment and to the former Russellian one, proceeds from an unsolved old problem (see my 1987a, 552 *ff*): the attempt of preserving the relational theory of judgment through the requirements of propositional functions, which remained in the substitutional theory. In his first exposition of this theory Russell stated clearly that in the new notation $p(x/a)$, destined to replace the functional one, p is the prototype and a the origin or *initial subject* because 'a may be taken as being, in a generalized sense, the subject of p' (1906a, 155).

In fact this was the central idea of substitutional theory, to start from an ontological pattern where only two kinds of entities were handled: simple (the constituents of propositions) and complex (the proposition itself), although this idea was also the one leading to the final destruction of the proposition as an entity (see Sect. 1.7 below). The goal now was to admit functions, which have the same subject predicate pattern (φx or φa, if we admit x or a as subjects), for semantic paradoxes force us to

[17] Landini claims to me that he has found new archival evidence to his original interpretation. In particular he says that *m1906b*, fols. 72 *ff* contain this textual evidence. I do not see at this place the conflict to which he points, but I hope to read his forthcoming paper on this matter.

regard 'classes of propositions' and these are characterized by the variations on the predicates (functional-occurrences) and not on the subjects (argument-occurrences). Thus, it is necessary to consider expressions such as 'all values of φ' which leads us to speak about a 'functional substitution' where p/a transforms into $\varphi y/a$, $p(x/a)!q$ into $(\varphi y)(x/a)!(\psi z)$ and, in general, p in φx. And with that we should be back to the situation which tries to avoid substitutional theory (see 1906a, 155). Again, the admission of this functional substitution requires to give a meaning to expressions of the form $\psi z/\varphi x$, in the same way that in substitution of matrices, or to regard them as primitive ideas (m1906a, fols. 2 *ff*).

But functional substitution presents problems in both cases. If we start from it as a primitive idea (Pi), we must admit always new functions of functions. Thus, if $\psi z/\varphi x$ is a Pi, then so are the two expressions $g(\psi z)/f(\varphi x)$ and $G\{g(\varphi z)\}/F\{f(\varphi x)\}$, which produces as a consequence an infinite hierarchy of Pi; and 'this is intolerable' (m1906a, fols. 10 *ff*). Yet if we start from propositional substitution, we can arrive easily at a functional substitution by means of $q/(p, a)$ (a class of classes, as we saw above). But here our notation conceals that we are referring to an expression of the form $f(\varphi x)$ for which it is necessary to find a meaning (fol. 23), and so we are led to the 'general form' of proposition and the vicious-circle principle.

However, if we start from $f(\varphi x)$ (φx to simplify) as a Pi, then we are neglecting propositions with apparent variables, as $(x).\varphi x$ (fol. 24). If, on the other hand, we take as Pi both expressions, the second one leads to contradictions when $(x).\varphi x$ being one of the values of x, or when we regard $(\varphi).\varphi x$ of the form φx. Just this possibility leads to the liar and the need of a hierarchy of propositions, with the only difference that here we speak about 'classes of propositions'. Thus, if φp has a meaning, it must be a proposition of a higher type than p, and we can eliminate expressions as $(\varphi)\varphi x$ (fol. 27). All that led Russell to replace propositions for functions (fol. 37):

> In fact, φx and $(x),\varphi x$ are homogeneous as to type; $\varphi(\varphi x)$ is just as shocking as $\varphi((x).\varphi x)$. There is *not* a hierarchy of types of propositions. The necessity for thinking so went with admitting $f(p\ ,a)$ or $f(\varphi, x)$. If *every* function has *only* entities as arguments, the hierarchy of propositions vanishes.

With that Russell seems to maintain the substitutional theory, but he claims immediately that we find the same problem when we quantify over propositions; as, for instance, when we formulate the law of excluded middle in the form '$(p). - p\text{V}p$', and the only way of avoiding it would be to specify the propositions we are referring to; when this device is not possible we arrive also to fx (fol. 38). Finally, we are told that the apparent variable does not work with propositions, but only starting from φx. The argument seems now definitive (fols. 218–219):

> None of the things we say start from propositions, but all from functions. For we always have a variable in our propositions, and we assert something about 'any value', 'all values' or 'some value'. Thus φx or (C)(α) is really our starting-point, not p.

However, the problems that the functional point of view produces in the substitutional device point out, again, to the limitations of substitutional theory in itself. Thus, if we consider φx and $(x).\varphi x$ as expressions of different types (and the same with $f(\varphi,x)$ and $(\varphi).f(\varphi,x)$), we are forced to admit new types until the infinite, with which we obtain new Pi of the form $F\{(\varphi).f(\varphi.x)\}$ (fol. 58). Again, 'the essence of contradiction consists in regarding $(\varphi).f(\varphi.x)$ as a case of φx', and so the infinite hierarchy of Pi's cannot be avoided, 'unless we keep propositions' (fols. 68–69). And, as we saw above, the return to propositions is not possible if we want to give an account of semantic paradoxes, unless we resort to a 'trick' (m1906a, fols. 73?, 80? – doubtful numbers), which had appeared many pages before and rejected, presumably because it presupposes the old theory of types: to avoid, by a principle, that $(x).\varphi x$ can be a possible argument to φ (and the same for more variables).

At this point we already obtain the three basic elements of 1906c. Firstly, the application of the mentioned principle to propositions: 'When a proposition is asserted for all values of a variable, the proposition itself must not be among the possible values of the variable', and this limitation must operate rather on 'determination' of a variable for a constant than on 'substitution' of a constant for another one (m1906a, fols. 31–33). In 1906c the principle was formulated in terms of apparent variables, but in the present form it allows the distinction between propositional and individual substitution, with which the theory can be put at work: 'the different sorts of substitution may be indicated in the same way: q/p is a propositional substitution, x/a an entity-substitution'. Now we can reject the substitution of a proposition for an entity (and the converse): 'q/p and x/a are to be meaningless' (m1906a, fol. 94), which leads to the general rule according to which 'a proposition must not replace an entity', and this rule applies against both substitution p/a and functional substitution $\varphi p/\varphi a$ (fol. 131).

Secondly, there is the distinction between propositions and entities. According to the stated rule, in matrices of the form $p(x/a)!q$ it would be necessary to know if x or a are propositions, before the possibility that the matrix $p(s/r)!q$ is to be meaningless, always depending 'upon the nature of p, q, s, r' (fol. 131). That is to say, in the same way that in some expressions as '—p' or '$p \vee q$' it does not matter what kind of propositions are contained therein, there are others, such as $p(x/a)!q$, in which that is fundamental. Summing up, '$p(x/a)!q$ is *not* a truth-function of p' (fol. 132). This second element allows (with the help of the principle mentioned above) to make use of functional substitution with no danger of paradoxes (fol. 135).[18]

[18] Of course, it is obvious that in 1906c 'there is really nothing that is not an individual' (p. 206), with which propositions would be as well individuals (at last they must be entities, or facts: p. 207). *But* there are still two problems: Russell speaks about 'general propositions' and maintains a distinction between proposition and 'constituent'. Thus, we must admit a kind of non-propositions that, as mere *façon de parler*, Russell calls statements, which suppose the first seed of the later ontological attack against propositions. On the other hand, the distinction between proposition and constituent, plainly admitted in 1906c, 200, is the first requirement of substitutional theory. But this requirement contained the danger (pointed out for the first time by Grattan-Guinness 1974a as another difficulty of substitutional theory), which was examined mainly in unpublished manuscripts, of substituting a proposition (p) for a constituent (a) and vice versa, which would lead (obviously) to nonsenses (see section 1.4 above). Thus, although I use here the words 'proposition-entity' (in the same way than Russell in the manuscripts), the important thing I am emphasizing is really the distinction 'proposition-constituent', in the sense that propositions are to be regarded as complexes. As I shall conclude, that was interesting as a general attempt of stating a relationship between the distinction

Lastly, we have the distinction between statement and proposition. Until this point, the theory involves the hierarchy of propositions, but requires that propositions stating the limitations do not violate these limitations: φx must not be interpreted as involving 'all propositions', which is a meaningless expression (m1906a, fol. 162). In other words, φx is referring only to some of the expressions which are functions of x; this is the only way of avoiding 'all values'. But then to define φx becomes impossible: 'This operation demands that the variable shall confine itself to one type. The difficulty is to get a non-circular definition of φx' (fol. 169), and to overcome it Russell chooses a version like that in 1906c (m1906a, fol. 171; the emphasis is mine):

> We must be able to explain what are meant by φx without mentioning in the explanation that 'all values of φ' are to be excluded; otherwise our explanation is circular. The explanation must be such that 'all propositions of the form φx affirmed by A are false' is not of the form φA. I think we shall have to say simply: φx is any *statement* about x; and a statement about 'any statement about x' is not a statement about x.

Now we can see how the distinction between proposition and entity contained the seed of the distinction between proposition and statement. Thus, the three basic elements of the published theory (1906c) can be explained altogether by means of an unique line of evolution in m1906a, which is not so clear in the published version, because it was mainly the exposition of the *results* obtained in manuscripts. The liar, as the supreme representation of semantic paradoxes, produced, therefore, doubt over the ontological status of propositions as entities: as his paradox of classes led to the elimination of classes, so the liar paradox of propositions led to the elimination of propositions (at least of some of them). Hence we must alter our presentation of the 'simple' substitutional theory as a return to the Moorean ontology of simple and complex terms. Semantic paradoxes force the distinction between these two kinds of terms, making impossible for propositions to be in the same ontological level than 'entities'. Thus, the distinction between individuals (logical subjects) and propositions makes easier the later distinction between propositions being genuine individuals and 'propositions' not being so (statements). As we shall see, the consequence of these problems was the functional view, and a theory of judgment which allows to dispense definitively with propositions, though these two results are completely clear only in *Principia mathematica* (1910a).

1.6 The Manuscript 'The Paradox of the Liar'

Resorting to m1906a, we have explained the actual historical origins of 1906c. But at this point Russell still must undertake three tasks: (i) to infer all consequences from the new modifications; (ii) to consider in detail the advantages and difficulties of, respectively, matrices, propositions and propositional functions, and (iii) to make

---

between proposition and constituent (entity), and the distinction between proposition and statement, which led finally to the abandonment of the old Moorean ontology presupposed in the 'simple' substitutional theory.

a valuation of the effects that the changes produced in the old theories of truth and judgment (for this evolution, see my 1987a, 55 *ff*, and 1988e). The manuscript 'The paradox of the liar' (m1906b), from September 1906 (and corrections from June 1907), was devoted to these tasks, as the transition between substitutional theory and theory of types.[19] As so many manuscripts, it is a complex set of possible lines and developments, many of them only suggested; but, unlike m1906a, it has a certain continuity. I shall try to extract only the relevant ideas to understand the logical and ontological evolution of Russell between 1906 and the last version of theory of logical types in 1910.

The first problem was the 'nature' of proposition, which underlies all the manuscripts. The point of view is, again, ontological: if a proposition about all propositions cannot be a member of this same class, we must reject it, and not only because of paradoxes but also because 'we must find some reason in the nature of propositions which shows that the impossibility subsists' (*m1906b,* fol. 2). After that, the several possible hierarchies are examined.

The first philosophical basis of the hierarchy is to reject variable proposition (*m1906b,* fols. 16 *ff*), using only entities combining among themselves to produce elementary propositions, which, by means of disjunction, negation and variation (to transform a constituent into a variable and to generalize), produce the hierarchy. But to formulate here the laws on deduction we must involve 'any proposition', so we must discard the hypothesis (fol. 22).

Another philosophical basis is the admission of propositions, but without apparent variables, which should be contained in 'statements'. It was the possibility actually used in 1906c (see above) but it has the difficulty of requiring an axiom of reducibility, which seems to Russell 'highly undesirable' (m1906b, fol. 25). Russell then tries a new way of dispensing with apparent variables: to add new Pi's. But finally he concludes accepting classes and relations as entities, so avoiding an axiom of reducibility (classes themselves produce reducibility). Thus 'simple' types avoid contradictions and the advantages of the statement-proposition distinction hold (fol. 29).

The hierarchy must be constructed here according to the number of apparent variables: genuine propositions lack them, whereas statements containing them must be regarded as ambiguous assertions of propositions; by this reason 'a statement about all propositions does not state a new proposition' (m1906b, fol. 30). Reducibility is obtained by means of classes themselves: 'Given any statement φx containing any number of apparent variables, there is, we may suppose, a class u, such that φx is formally equivalent to $x \in u$' and this expression contains no apparent variable (fol. 31). In this way the old problems about classes arise again, but they can be avoided by logical types: individuals, classes, classes of classes, etc. However, the problems leading again to the functional point of view are to appear soon.

[19] There are two other manuscripts related which I have not seen, but it seems they are fragmentary and brief. They are 'Logic in which propositions are not entities' (April-May 1906) and 'Fundamentals' (January 1907). At any rate m1906b has besides the interest of containing corrections from June 1907, a date very close to the writing of 1908a. Apart from these, there are more material on substitutions; I. Grattan-Guinness tells me that in 1984 he reconstructed one of 262 folios.

Firstly, this theory eliminates apparent variables: with 'all classes' we obtain the same as with 'all propositions'. Thus, statements containing apparent variables can be regarded as genuine propositions: it suffices to forbid that a proposition can be an apparent variable in itself (m1906b, fol. 37). Now quantification over propositions results eliminated, and the hierarchy of propositions (and functions) disappears, despite of propositions losing their status of entities when replaced by classes (fols. 39–40).

Secondly, we must regard classes as non-individuals; otherwise 'a class could not, without a vicious circle, be defined by any function in which an individual appears as apparent variable'. And this can be made only with some axiom of reducibility, according to which 'every proposition about a class a is to be of the form $f!(\hat{x} \in a)$', where f may contain any number of apparent variables, a is the only real variable, and every propositional constituent containing a is of the form $x \in a$ (x being a constant or an apparent variable). The axiom must assure that, no matter how many apparent variables are contained in f, the form mentioned is equivalent to one containing only entities and classes as apparent variables (*m1906b,* fol. 46):

$$(\exists\varphi):f!(\hat{x}\in\alpha).\equiv_x.\varphi(\hat{x}\in\alpha)\ \mathrm{Pp}.$$

But this introduces the problems of the old axiom of reducibility; so Russell concludes: it seems the hierarchy of functions is more intelligible when we don't introduce classes etc. than when we do' (fol. 48).

Thirdly, the admission of classes should lead to these two Pp's:

$$(\exists\alpha) : \varphi \mathrm{x}. \equiv_\alpha .x \in \alpha \quad \text{and} \quad (\exists\beta) : \psi\alpha. \equiv_\alpha .\alpha \in \beta,$$

where the conditions of significance of $\psi\alpha$ are not clear. But this should require modifications; for instance, to declare false (rather than meaningless) the expression '$\alpha \in \alpha$'. According to Russell this is possible, but introduces 'obscurity and arbitrariness' in the development. Thus, the method by functions is more 'natural and straightforward', whereas its only difficulty, the axiom of reducibility, can be very much deferred (*m1906b,* fol. 51).

The final list of Pi and Pp (fols. *55 ff*) is very similar to the one published in 1908a, although a little more complex. An ontologically important feature is that the first Pi is 'any proposition', which disappears in 1908a, once he admitted the new theory of judgment (see below). Even here, some pages after that, while trying to explain in a simpler way the resulting hierarchy, Russell eliminates this idea, giving so the final pre-eminence to propositional functions, as appears in 1908a and 1910a (although still considering classes rather than individuals): 'The first *Pi,* that of "any proposition", is not strictly necessary; for we may write *fa* in place of *p*' (fol. 59; also in fol. 93). Where we must understand *fa* as any proposition about some or all values of *a,* and this inside a hierarchy of functions in which only arguments are relevant.

However, propositions cannot be easily rejected. When Russell points out the ontological problem, he admits that an individual is only 'any single existent' (fol.

64), but he must admit some being for propositions because they can produce apparent variables (a harmless being with the hierarchy of types). But when we so proceed, the problem of the constituents appears: to deal with them we need the notion of substitution (fol. 68). At this point, Russell attempts a general valuation of substitutional theory ending in its rejection. However, the difficulties appear only in the manuscript, whereas in 1908a the theory is rejected unclearly (as we shall see below).

Russell mentions many advantages of substitutional theory as well, but some of them in a somewhat unclear way. Firstly, it is a great philosophical merit that substitution can produce a distinction between a propositional function containing x and another one expressed in the form φx, where φ may be variable (fol. 70). This is necessarily true because the expression $p(x/a)$ transforms only the constituents of the proposition into variables. Secondly, substitutional theory admits propositions, 'though not in the same sense in which there are individuals' (fol. 72), which produces the difficulty of the distinction between entities and propositions. Besides, the substitutional theory: (i) permits to state classes of propositions starting from orders[20] (while logical types avoid substitution of a proposition for an entity); (ii) avoids the need of functions of two or more variables as Pi for each number of variables (because of the avoiding of self-application of functions); and (iii) makes easier to obtain an axiom of infinity because it permits us to speak about *constituents* of propositions (fols. 71 and 73).

But there appear two objections which seem to be definitive. The first one arises when Russell denies his former claim according to which functions, as apparent variables, can be replaced by matrices (fol. 66), which is just the possibility that he mentions in 1908a without developing it (see below). Russell writes that if φy is really a denoting expression, y being constant, the substitutional theory does not obtain: 'For it seems to demand that $\varphi\, \hat{y}$ should be taken into consideration, and φy regarded as $(\varphi\, \hat{y})y/\hat{y}$. But if $\varphi\, \hat{y}$ must in any case be admitted, the advantages of substitution are lost' (fol. 87). With this argument Russell seems to mean that in that case it would be not necessary to use propositions, because he adds that the notation φy is an instance of $\varphi\, \hat{y}$, and, therefore, 'approaching it by a denoting expression. Thus, 'if we put down the actual value which $\varphi\, \hat{y}$ has in the case of the argument y, φ has been swallowed up' (*ibidem*).[21]

The second objection, also difficult to follow, takes place when Russell, once convinced about the pre-eminence of functions over propositions, tries to apply substitution to them. Thus, instead of $p(x/a)$, we should have *the expression* $(\varphi b/a)x$, that is to say, the propositional function resulting from substituting b for a in φx.

[20] At least here, Russell shows he sometimes distinguishes clearly between use and mention: 'In a 1st. order proposition, though we *assert* a proposition we don't *mention* one; in a 2nd. order proposition we mention a proposition of the 1st. order. Thus $p(x/a)!q$ is a 2nd. order proposition, but *p**q* is not' (*m1906b*, fol. 71).

[21] In June 1907 he added a note expressing his doubts about this objection: '*x denotes q*, where q is the actual value for the argument y. If p is the value for the argument a, $p(y/a)!q$. The point wants reconsidering' (*m1906b*, fol. 87). However, it seems that the reconsideration was never made, as is shown by the fact of the very abandonment of the theory, which was supported, as we are seeing, by other reasons.

However, the need to give an account of expressions like $f!\varphi\,\hat{z}$ leads to admit $(f\chi/\theta)!\varphi\,\hat{z}$ and, consequently, Pi and Pp become very complicated. This seems to finish with the advantages of the theory, 'for the chief point of substitution was to avoid functions as apparent variables, and this plan does not do that' (fol. 105).

The only way out seems to be, therefore, the functional point of view, as followed in 1908a and 1910a. In *m1900b* the corresponding hierarchy seems to be the soundest attempt, although it is entitled as 'the propositional hierarchy'. It starts from individuals and predicates (and relations) as 'the sum-total of the simple constituents of the universe'. After that, we have certain *elementary* propositional functions with only individuals as arguments. Applying now the processes of negation, disjunction and variation (see above) to these functions, we obtain new propositional functions called *properties of individuals,* whose values for the argument x are denoted by φx. The new variable φ stands for 'any property of individuals'. The repetition of these processes leads to new propositional functions, to *properties of classes,* and so on. Statements about functions of any order are treated as statements about their *values* (*m1906b,* fol. 96).

Finally, 'The paradox of the liar' throws light as well on the logical origins of the 'multiple' theory of judgment, with which the changes between 1906 and 1908–1910 are much better understood. In his earlier work Russell came again upon the usual ontological difficulties concerning functions and propositions. For, if a function is only an 'ambiguous proposition', then it has no kind of being apart from the being of its values (*m1906b,* fol. 97); unless we admit that ambiguity itself has some kind of being, which seems undoubted (fol. 106). However, the being of an ambiguity consists only in the being of its values, which are just propositions not containing apparent variables. Therefore, 'the ultimate values of any function are always propositions which contain no apparent variables. Thus these alone need have being' (fol. 107). And, as such values are always of the form φx where φ and x are constant, these will be the definitive constituents which must be regarded by processes of variation as forming different combinations (fol. 108). And this supposes returning to the old view of propositions as complexes composed of constituents whose basic combination is that of subject predicate (see my 1987a, 552 *ff*).

Thus, despite the pre-eminence of the functional point of view, propositions cannot be completely discarded. However, when we try to give an account of propositions, we are forced to analyze them into their components. This ambiguity makes the modification of the old theory of judgment, according to which the proposition was a single entity (a logical subject), unavoidable, until the theory finally published in 1910 after some doubts (see my 1988e): judgment is a 'multiple relation' between mind and the several constituents of propositions (with the corollary about the non-existence of false propositions). But this new theory is already present in m1906b, which is an undoubted sign that it was a *consequence* of logical and ontological problems concerning definability, and not a *cause* of them (see Sect. 1.5 above).

In fact, the distinction between individuals (entities, constituents) and propositions (combinations) makes that judgment (here *belief*) cannot be a single entity. This is the more relevant passage (*m1906b,* fol. 63):

It is necessary to the above theory to hold that a *belief,* unless at best when no apparent variables occur, is not a single state of mind, but a complex, presumably involving transition. For if there is a belief for each proposition, then, since beliefs must be individuals, there will be at least as many individuals as propositions, which is impossible. Hence a belief in a complicated proposition is not to be regarded as one state of mind, but as a sequence or coexistence of several.

Russell's argument, which seems inspired by Dedekind (see my 1987a, 487 *ff*, and 1988d), is clear: the distinction between individual and proposition (made necessary by logical motives) leads to regard judgment rather as a multiple relation between mind and each constituent of proposition (individuals), than as a unique relation between mind and one object (proposition). Propositions must be regarded as combinations of individuals and, therefore, more numerous than those; otherwise, also propositions should be individuals (as in primitive Moorean ontology; see my 1988j). This seems to be the main argument which led Russell actually to the new theory of judgment *already in 1906.*[22]

This modification was already needed in substitutional theory, because this theory eliminated the Moorean non-distinction between simple and complex terms; but Russell did not make it until he faced in detail the ontological problems of propositions in connection with the looking for a definitive hierarchy of logical types. The same consequence concerning false beliefs was inferred many pages before that (m1906b. fol. 4):

The first difficulty in such a view [that only true propositions subsist objectively] is to explain what we *mean* by belief in what is false. In order to explain this we shall have to regard what is called belief in a proposition as not a thought related to the *proposition,* but rather a thought related to the *constituents* of the proposition. I.e. if I believe that A is B I have thought related to A and to B, but not to the proposition 'A is B'.

Therefore the distinction between individual and proposition (needed also by *logical* reasons in substitutional theory) made impossible the old theory of judgment, even without the former technical argument.

Consequently, we can accept the original insight by Grattan-Guinness 1974a, according to which the new theory of judgment led finally to the abandonment of the substitutional theory, but with an important correction: the admission of the new theory of judgment (already considered in 1904a) took place in 1906, and it was a

[22] I agree with I. Grattan-Guinness when he tells me that the first date in which the new theory appears is a 'chronological nightmare', for even in 1904a Russell had been considering several different judgment theories. In fact we can find a very similar theory in 1904a, but the difference consists in that, whereas there he considered the new theory only as a possibility which must be rejected, in 1906c he admits it. The relevant data from *1904a* are the following: the nexus between complexes and propositions (pp. 50 *ff*); the idea that complexes cannot be thought apart from their terms (p. 27); the view of false propositions as non-existent relations (pp. 38, 48-49); the differences between our attitudes before simples and complexes (pp. 62-63), and finally almost the exact formulation of the theory of judgment as a multiple relation (pp. 69-70). Curiously, the theory is rejected because Russell regarded as unacceptable that, 'if it were correct, we could never become aware of complexes'. It is a feature of Russell that he sometimes considered in detail and rejected theories that he adopted later (other examples: the theory of transfinites by Cantor; the definability of descriptions; and neutral monism).

consequence of something presupposed in the substitutional theory (the distinction between proposition and constituent), which led as well to other additional problems. Thus, the acceptation of the new theory was not an independent decision motived only by general philosophical problems, but in particular by the difficulties of the ontology of propositions in the context of semantic paradoxes, and doubtless facilitated by the eliminative method which previously had led to dispense with descriptions and classes.

Summing up: the problems of substitutional theory led to its abandonment, but the alternative of classes had also its own objections, which led finally to functions. But the functional point of view, although preferable to a pure propositional hierarchy, required a certain admission of the being of propositions, which, again, produced unusual and dangerous epistemological consequences. Thus, propositions must be eliminated; however, this was impossible on pain of emptying the functional view of ontological content. In 1908a Russell chose, following Peano, to juggle with the ontological status of propositions: they are neither a primitive idea nor a derivative one (a situation somewhat similar to that of *Principles).* As we shall see, the solution in *Principia* was very drastic: to declare as well propositions as incomplete symbols (although introducing elementary propositions as a Pi), with all the consequences that this had for the eliminative resources.

1.7 The Final Theory of Types as an Ontological Engagement

I shall expose firstly the basic problems of the final theory of types (as it was published), especially the ontological ones. In that way the manuscripts we have studied will contribute to explain: (i) why the last version was chosen; (ii) what was the significance of the differences between versions of 1908 and 1910, and (iii) how their difficulties can be interpreted. From the point of view of constructive definition (see Sect. 1.2 above), the central ideas of 1908a are the following.

First of all Russell solves the problem of reflexive fallacies relative to 'all values' of a variable by means of the distinction between 'all' and 'any', that is to say, by means of the distinction between apparent and real variables. Deduction requires real variables, but definitions require apparent ones, so we cannot omit the second kind in favor of the first one. Russell chose the way to distinguish between generalized propositions and propositional functions. The first ones contain the word 'all' (or 'some'), whereas the second ones contain a real variable (1908a, 69). In doing so, we can transform a categorical proposition as '$(x).\varphi x$' into the affirmation of a propositional function for all its values: 'φx always', making the restriction of the variable an *internal* property.

In that way the arguments for which the function is not true result automatically outside of its *range of significance,* without the need of explicit statement (which should produce the same problems), because 'it is impossible for a function to be true more generally than for all its values' (p. 72). Thus, 'if x is a man, x is mortal' is always true even when x is not a man (avoiding so the restriction of the variable within a certain range), just because the propositional function in which the variable occurs remains significant when the variable is outside the range' (p. 73). The pre-eminence of propositional function is therefore definitive (p. 75):

> Every proposition containing *all* asserts that some propositional function is always true, and this means that all values of the said function are true, not that the function is true for all arguments, since there was arguments for which any given function is meaningless.

At this point the hierarchy of types appears: these are only the ranges of significance of propositional functions (rather of the values of their apparent variables). The mere division of objects into types solves immediately the reflexive paradoxes, because whatever contains an apparent variable will be of a higher type than the values of this variable. But propositions can be values also, as in semantic paradoxes. For this case Russell introduces a hierarchy of propositions, which starts from *individuals* (entities without complexity) and *elementary propositions* (whose *terms* are individuals). Starting from these, we can obtain *generalized* propositions (containing some apparent variable) by replacing one of the terms of an elementary proposition for a variable and applying universal quantification (1908a, 75). In the new propositions so obtained, the first logical type is that of individuals; the second one is elementary propositions together with *first-order* propositions (whose apparent variables are to be individuals); the third one is *second-order* propositions (whose apparent variables are to be first-order propositions), and so on. In general: 'The $n + 1$th logical type will consist of propositions of order n, which will be such as contain propositions of order $n - 1$, but of no higher order, as apparent variables' (pp. 76–77).[23] This solves the liar paradox whose assertion must be referred necessarily (unless to be meaningless) to a determined order of propositions.

But Russell prefers a hierarchy *of functions*. The method of substitution allows him to obtain *matrices* (p. 77):

> If p is a proposition, and a a constituent of p, let '$p/a;x$' denote the proposition which results from substituting x for a wherever a occurs in p. Then p/a, which we will call a *matrix*,

[23] The distinction between types and orders seems not to be completely clear. In the first approach it seems that types are referred to objects and entities and orders to propositions and functions. For if propositions (or functions) are taken as entities, then types must affect them. This situation has produced two main interpretations among commentators: the more traditional one, which regards orders as a further division of types, surely because this was the order in which these two categories were introduced (see for instance Kleene 1952a and Kneale and Kneale 1962a); and a second one which regards orders as divided into types under the argument that, for example, in order 2 we can distinguish two classes of propositional functions: those taking individuals as arguments and those taking propositional functions of order 1 as arguments, with which the order 2 results divided in two types, order 3 in 4 types, etc. This version (from Chihara 1972a and 1973a and Bunn 1980a) seems the more correct one. Anyway, the distinction from Ramsey remains relevant: 'The type of a function is a real characteristic of it depending on the arguments it can take; but the order of a proposition or function is not a real characteristic, but what Peano called a pseudo-function. [...] Order is only a characteristic of a particular symbol which is an instance of the proposition or function' (1931a, 47). See also Copi 1971a, 84 *ff*.

may take the place of a function; its value for the argument x is *p/a;x,* and its value for the argument a is p.

But although this device could allow apparent variables to be omitted and introduced 'a certain simplicity' into types, Russell describes it as 'technically inconvenient' and chooses the replacement of p for φa and of *p/a;x* for φx; that is to say, he goes back to *propositional functions,* just in the contrary way started in 1906b and modified in 1906c.

The role of φ as apparent variable is stated by a hierarchy of functions. First-order functions will have individuals as arguments and first-order propositions as values; second-order ones will have functions or first-order propositions as apparent variables, and so on. In terms *of predicative* functions (whose order is the one immediately higher than those of its arguments and whose notation is $\varphi!x$) it will be possible to transform φ into an apparent variable: first-order ones cannot have functions as apparent variables, but individuals; second-order ones have second-order propositions as values and first-order functions as arguments (f in '$f!(\psi!x)$'); third-order ones have third-order propositions as values and second-order predicative functions as arguments; etc. (1908a, 78). All that solves paradoxes.

The usual *axiom of reducibility* allows to handle propositions including 'all properties of x', which are necessary in mathematical induction. Its main idea is the reduction of the order of propositional functions, which is equivalent to regard classes as individuals: 'every propositional function is equivalent, for all its values, to some predicative function' (p. 81). This makes possible, for higher types of arguments, that predicative functions take the same place that first-order functions in respect of individuals. Also, it suffices to state mathematical induction, for instance, for all predicative functions of numbers; that is to say, for *any* function of whatever order (p. 82). In symbols (p. 87):

$$(\exists f) \therefore (x) : \varphi x . \equiv . f!x.$$

That means, ontologically, the abandonment of classes. Firstly, reducing them to individuals which are to be used as the arguments for certain functions, by means of the notation $\hat{z}\ (\psi z)$, similar to the class abstractor from Peano and the *Wertverlauf* from Frege. Secondly, defining contextually such 'fictitious objects' (or 'incomplete symbols') with a method very similar to the one applied in eliminating descriptions (p. 89):

$$f\{\hat{z}(\psi x)\}. = :(\exists\varphi):\varphi!x . \equiv_x . \psi x : f\{\varphi!\hat{z}\} \quad Df.$$

That is to say, eliminating classes (and relations) in terms of propositional functions, which are introduced as *primitive ideas* (p. 83). We shall not follow Russell in the construction of cardinals and ordinals, which is outside my present purpose (see my 1987a, 467 *ff*, and *1988d*). Concerning paradoxes, Burali-Forti's one is solved by dividing ordinals into types, which makes to speak about 'all' them impossible.

Unfortunately, Russell eludes all philosophical difficulties implicit in the former general construction, appealing to his known distinction between mathematics and philosophy; according to him, such problems are to be 'separated' from the mathematical development of the theory, because they 'introduce elements of uncertainty which do not belong to the theory itself' (1908a, 102). However, the most interesting thing for us is just to make explicit such difficulties; for that, I shall compare the ontology used explicitly here with the one underlying substitutional theory, now depending on the insurmountable difficulties as stated in the manuscripts.

The main differences between both theories concern the respective status of propositions, propositional functions and matrices. As for propositions, in 1908a they are admitted as entities in general (with the exception referred below), mainly in the new hierarchy of propositions, in which we are told it is possible to regard them as values of apparent variables. However, they are accurately distinguished from individuals (which are deprived of complexity) with the aim of making easier the construction of the corresponding hierarchy by means of the process of generalization, which starts from individuals (1908a, 76). It leads to exclude, apparently, the distinction of 1906 between statement and proposition, which was destined to reject as non-genuine the propositions containing apparent variables. However, the spirit of that difference is maintained through the one existing between elementary and generalized propositions, and 'statements' are preserved to design (so it seems) the corresponding linguistic expressions.

The second change concerns the restriction of apparent variables, which, as we saw, is now an 'internal' limitation, without new assumptions stating so explicitly are needed. However, matrices appear, although only to point out the (technically inconvenient) possibility of replacing functions. All that represents a clear pre-eminence of propositional functions, which are admitted as primitive ideas and are used to eliminate classes and relations. Russell does not mention the technical and ontological difficulties of the substitutional device, as stated in the manuscripts, but I think they underlie the 'technical' inconvenience.

Therefore, it is not possible to agree with Landini (1987a, 190) who holds that in 1908a propositional functions are not entities, but mere convenient subjects which 'disappear upon analysis'. In holding this view, Landini follows Cocchiarella (1980a, 97) who maintained a similar assimilation of propositional functions to classes in 1908a. And he does so by appealing to the same argument: because the possibility pointed out by Russell according to which functions are to be replaced for matrices, which could suppose their elimination in favor of propositions and their constituents. However, Russell rejects finally this possibility and maintains propositional functions even as *primitive* ideas (namely, irreducible and indefinable inside the system), doubtless because of the reasons appeared partially in m1906a and completely, as we saw, in m1906b.

This pre-eminence of functions over propositions seems to throw some doubt on the status of these last entities. Landini (1987a, 188), again following Cocchiarella (1980a, 95), insists in the ontological character of propositions, but mainly with the support of this passage (apart of the hierarchy mentioned above): 'Whatever we suppose to be the totality of propositions, statements about this totality generate new propositions which, on pain of contradiction, must lie outside the totality' (1908a, 62). But it seems clear that Russell rejects only any mention of *all* propositions, and so he is referring to a determined range of significance. A sign in favor of this could be that in another place he writes: 'These laws [of logic] have, so to speak, a particular enunciation but no general enunciation. There is no one proposition which *is* the law of contradictions (say); there are only the various instances of the law' (1908a, 68: see also p. 62). It is usual to point out the lack of metalogic distinctions in Russell, but anyway these passages seem to show that Russell did not abandon completely his earlier distinction between propositions and statements, with which the ontological hierarchy of propositions is doubtful.[24]

Besides, Russell seems to reduce in some places the former quantification over propositions, necessary in the corresponding hierarchy, to the concept of propositional function: 'the fundamental *all* is "all values of a propositional function"'; 'Every proposition containing *all* asserts that some propositional function is always true' (1908a, 72, 74–75). With that, the whole matter seems to be led directly to the functional hierarchy as the underlying basis of the apparent propositional one (explicitly in *Principia*). As explained above, I think that the new (and already accepted) theory of judgment was implicitly present here, as we can see, for instance, in the fact that Russell supports all the propositional hierarchy *just with the distinction between individuals and proposition* ('no individuals should be propositions', p. 76), against the view from 1906c. But this distinction had led in m1906b (and later in the substitutional theory of 1906c) just to the ruin of the ontological status of propositions and, consequently, to a very strong argument against the technique of matrices. Therefore it seems that the admission of propositional functions was not by mere convenience, but simply the better recourse to obtain a presentation of the theory of types which was (by the moment) free of the philosophical labyrinth of the status of propositions.[25]

---

[24] Landini follows Cocchiarella in pointing out this passage as the better ground for a possible status of propositions as logical subjects in Russell 1908a. But unfortunately he does not follow Cocchiarella in pointing out that Russell avoids any discussion 'of the distinction or similarity between statements and propositions, especially as regards the sense in which the later may also be said to contain an apparent variable' (Cocchiarella 1980a, 113). Thus, we must not forget that *generalized* propositions are dangerously similar to *general* propositions from 1906c, which were mere *façons de parler.*

[25] Of course, limitations of the range of apparent propositional variable in Russell 1908a are related to orders of propositions, but the very notion of *type* has a sense only in terms of the ranges of significance of propositional functions (p. 75), although we must introduce propositions because propositional functions presuppose them in some way. Landini (1987a, 195) has written: 'Russell does not begin with propositional functions as logical subjects and then attempt to show that every truth capturable in a language for them is capturable in the language of matrices. Rather, he starts with matrices (i.e. with propositions as logical subjects), and uses a language of propositional functions as a technical convenience'. It is true that in section IV of 1908a Russell begins with the hierarchy of propositions, *but:* (i) there is no point in this hierarchy without the notion of *type,* which was introduced in the first two lines of that section in terms of propositional functions; (ii) the *real* beginning of 1908a as a system is the list of Pi and Pp of section VI, and *there* Russell begins with propositional functions (Pi number 1; 1908a, 83) and introduces no idea of proposition, not even of elementary propositions; therefore (iii) propositional functions are *indefinable* inside the system, and any other notion introduced later must be defined in terms of them along with the

We must conclude that there is only a superficial mention of substitutional theory in 1908a (and no trace of it in *Principia*), which can be perfectly omitted without any danger for the working of the presented theory of types as a whole. This does not mean, however, that all the problems concerning propositional functions were solved, but only that these entities supplied the more acceptable way out of a very complicated situation. It is true that this kind of entities can play the role of classes and so to generate paradoxes (as Russell discovered in *Principles*), but with the corresponding hierarchy paradoxes seems to disappear. Therefore, although in 1908a Russell identifies, operatively, classes and propositional functions (at least in the *extensional* sense), the respective ontological status are very different. The first ones are nothing, but the second ones are *primitive ideas*, and even logical subjects, as shown in the quantification over them and the elimination of classes in terms of this kind of functions. It is true that Russell sometimes speaks after 1908 as if propositional functions must be regarded only as *expressions*, in the same way than classes (as for instance in 1959a, ch. 7), but we must remember that propositional functions generate not only classes by means of the variable, but also propositions, which are the values of the function. Thus the ontological status of propositional functions needs the admission of propositions, which leads to give *some* response to the problem of the nature of propositions (as Russell says many times in the manuscripts) and, consequently, to some theory of judgment.[26]

Landini has another argument against the apparent role of propositional functions in 1908a: he has suggested a way to state the axiom of reducibility, which was formulated by Russell quantifying over functions, avoiding any commitment to these entities (see his 1987a, 194). Under my interpretation of the whole matter it is obvious that there is no point in this attempt, no matter how skillful it is: Russell used propositional functions not only in 1908a, but even in many places of *1906c* (as explained above). However, there is an explicit argument against it. The suggestion

rest of Pi (as, for instance, matrices, which Russell does not mention after that). We must not forget that Russell was explicitly avoiding philosophical problems, and obviously the *American journal of mathematics* was not the best place for a difficult, and perhaps unfinished philosophical view about the new theory of judgment and the fictitious status of propositions, as it appeared in 1910a. I think that the strange status of propositions in 1908a was a first effect of the underlying new theory of judgment, whereas only in 1910a the true 'logical order' was faced by dispensing with propositions as entities, the abandonment of matricial techniques, and propositional hierarchy as derived from the functional one.

[26] Concerning the technical possibility of replacing functions by matrices, it is rejected by Cocchiarella (1980a, 99), who, however, admits that for Russell propositional functions are fundamental as logical subjects needed to eliminate classes. But he points out this problem because he thinks that Russell needed to eliminate propositional functions in terms of matrices (like Landini), which we saw is thoroughly open to doubt. Anyway he states an interesting problem on this possibility (1980a, 100), which he regards as admissible only in the case of categorical propositions requiring universal quantification over functions, but not in the case according to which a function appears as a single logical subject, as in '(*f*)', just the form appeared in the contextual elimination of classes (see above). But we must remember that a few pages before Russell wrote: 'a proposition beginning with "(*f*)" is, by definition, the negation of one beginning with "(*f*)"' (1908a, 87). However, we saw above that Russell had many doubts on this possibility.

by Landini, which incidentally does not explain accurately the recourse to 'predicative matrices', depends on a fundamental idea: quantification over propositions. It is true that Landini's axiom is sufficient to avoid commitments to propositional functions, but the price he pays is too expensive: *quantification over propositions transforms the axiom of reducibility into a statement,* therefore it cannot be a *proposition* (using the terminology of 1906c). It is true that there Russell claims for a general reconstruction of all logical first principles, which show the same flaw: they contain apparent variables (p. 205). But there is a difference: Russell explicitly writes that the (later named) axiom of reducibility is only an 'assumption' stated in terms of quantification over statements, and not over propositions (p. 212).

We must not forget that 1906c was only a rough draft, but not a true attempt to reconstruct foundations of mathematics (from the logicist point of view) in detail. This attempt should need the establishment of a series of Pi and Pp, and doubtless the *axiom* of reducibility should be one of these last primitive *propositions,* and not simply a statement. But this point can be perfectly inserted in my interpretation of the changes between 1906 and 1908 (and even contributing to improve it). In 1908a the distinction proposition-statement *(general* propositions, namely mere *façon de purler*) is replaced by the distinction between *elementary* and *generalized propositions* (although, as I pointed out above, the former distinction seems to appear sometimes). Thus, if we try to state the axiom of reducibility, following to Landini, in terms of quantification of propositions, we should obtain only a *generalized proposition.*

There is no special argument against that, for all propositions containing apparent variables are generalized propositions (1908a, 69), and the axiom of reducibility contains such an apparent variable as stated in 1908a (p. 87). *But* as I explained above, the status of propositions is not clear enough in 1908a (they are neither primitive nor derivative ideas): it seems that they were introduced only to face the semantic paradoxes but avoiding all philosophical problems. But propositional functions were introduced in 1908a as Pi (as well as in *1906d,* no matter the different notation in terms of the old 'denoting complex') and even as logical subjects, as just the axiom of reducibility shows in quantifying over them. The problem was that propositional hierarchy seemed to be required in spite of the necessity to clarify the ontological status of proposition in itself and, consequently, of avoiding quantification over them (which appears directly only in the exposition of the propositional hierarchy, with the doubts I pointed out above, and indirectly in some isolated passage). This is the reason that possibly led Russell to introduce propositions before propositional functions and to preserve in some way the recourse to matrices. But propositional functions were used already to eliminate classes, with which matrices lose their first reason of existence.

Summing up: the nature of propositions was very obscure in 1908a. The recourse of the criterion of containing apparent variables was useful as a means of stating the distinction between generalized propositions (which contain some apparent variable) from propositional functions, but, as we saw above, it does not suffice to overcome completely the old distinction between propositions and statements. Likewise, the radical separation between individuals and propositions means that these last 'entities' are presented as complexes, with which the old Moorean ontological assimilation between simple and complex concepts is not to be maintained. This treatment of propositions, still incipient in 1908a but coming from the difficulties of substitutional theory (in particular from the substitution of a proposition for an entity; see above), required, besides, a new theory of judgment (already accepted in the manuscripts; see above), which was underlying the exposition of 1908a and appeared clearly only in *Principia.*

1.8 Incomplete Symbols and the Continuity of the Method

In 1910a (that is, volume 1 of *Principia*) the situation was clarified: (i) propositional functions as logical subjects are maintained, even as 'single things' (1910a, 40); (ii) it is claimed explicitly that the propositional hierarchy can be derived from the functional one, and even that it is never required except for the solution of paradoxes (p. 55), with which the necessity of regarding propositions as values of apparent variables disappears (they will be values *of functions*); (iii) the axiom of reducibility disappears as a standard Pp (as in 1908a) and turns to be an 'assumption', in the same way as in *1906c*; but (iv) propositions are claimed to be incomplete symbols, with which (v) it is no more necessary to maintain the language of *generalized* propositions (these only presented as incomplete symbols—*façons de parler*—in 1906c); (vi) the propositional functional hierarchy beyond elementary propositions suffices; (vii) quantification over propositions must clearly be avoided; finally, (viii) the new theory of judgment, already accepted in 1906b and eluded in (but underlying) 1908a, is officially proclaimed, with which the admission of 'elementary propositions' as Pi constitutes no danger (even when it seems to be somewhat incoherent). Now we can compare the theory of types of 1908a with that of 1910a.

As similarities, we find: (i) the structure of Pi and Pp, at least in general; (ii) the admission of propositional functions as Pi (1910a, 92), which has not been sufficiently noted, especially when the allusions of Russell in 1908a on the possibility of substituting matrices for functions is literally interpreted, and (iii) the status of classes and relations, although in 1908a the expression 'incomplete symbol' was not yet used. But *differences* are more interesting. In 1910a there is no allusion about the possibility of making use of matrices as raw material to define functions. Contrarily, we are told about them now as certain very general functions (1910a, 50 *ff*). Therefore, though classes and relations continue dissolving themselves, in 1910a they do so in terms of propositional functions: hence there is no sign of the 'spirit' of substitutional theory (as claimed by Landini 1987a), because the idea of eliminating classes is earlier. However, when propositions are regarded in 1910a as incomplete symbols their status has changed, and this is a sign that they played still in 1908a a certain role respecting to the possibility of substitution.

In 1910a the role of propositions is different: they are introduced as Pi (p. 91), at least the elementary propositions; but the functional hierarchy is not derived from the propositional one, but vice versa (p. 55). Besides, quantifying over propositions is not possible (p. 129), at least in theory, although it is not clear that this goal was always reached. Even the status of individuals is different: in 1908a they are 'defined' merely as not being complex, whereas in 1910a they are a Pi (p. 132), surely because of the doubts concerning the ontology of propositions (which are lacking in 1908a). The new 'explanation' was: '*x* is "individual" if *x* is neither a proposition nor a function' (1910a, 132).

Finally, in 1910a the need of a new theory of judgment is admitted (pp. 43–44; also in *1910b*). As we saw, the reason of eluding the new theory in 1908a was not the lack of arguments (already in m1906b), but to avoid the philosophical consequences of the distinctions individual-proposition and statement-proposition. Unfortunately Russell, who was trying to minimize philosophical problems even in 1910a, did not seem disturbed by his presentation of the 'multiple' theory of judgment as a *cause* of his attack against propositions rather than as one of its actual *consequences.*

Landini 1987a has claimed that substitutional theory was not abandoned even in 1910a. But it is obvious that there were no propositions, and they are the necessary basis for this theory. Besides, only in 1910a the problems concerning the pre-eminence of propositional or functional hierarchies were solved, following the line started in the manuscripts from 1906 (as I explained above). Anyway, I see no problem in saying that propositions are the values of propositional functions (which *are* ambiguities, because ambiguities also have being); we must remember that Russell never explained in a satisfactory way how it is possible that propositions are incomplete symbols when they are taken independently from the act of judging. But although we must admit problems in the philosophical foundations of *Principia,* to deny the logical order of the introduced ideas is a very different thing, and there the propositional hierarchy is derived from the functional one, as Russell discovered as the better solution in the manuscripts (see Sects. 1.5 and 1.6 above). Consequently I think that Russell introduced propositions as values only as an intuitive way to explain the nature of propositional functions, and before the new theory of judgment was mentioned.[27]

But to transform propositions into incomplete symbols was just to add more ontological complications. From the first formulations of the eliminative element of his constructive method, Russell holds that the elimination occurs contextually, namely as a paraphrastical reconstruction of the expression containing 'false abstractions'. But to call 'incomplete' to the eliminated symbols seems pointing out other

[27] G. Landini points out several related problems to me: (i) Did Russell allow propositional functions to be terms of the multiple belief relation?; (ii) Are propositional functions to be constructed as complex universals, in spite of the view from 1911a according to which no particular is a constituent of a universal?, and (iii) What about Russellian descriptions (for instance in 1913a) of propositional functions as expressions whose values are sentences? The first one is a very difficult question (which to my knowledge Russell never faced) and it depends on a previous one: what is the relationship between propositional functions and individuals? In principle I see no objection against Cocchiarella's treatment of the matter in his 1980a, but this question (as well as the second one) is not a problem I consider here. As for 1913a, Russell explained intuitively propositional functions, in exactly the same way as in *Principia,* as 'expressions containing a variable [...] and becoming a proposition as soon as a definite value is assigned to the variable'. Anyway, the new theory of judgment had a cruel destiny, due to the criticisms by Wittgenstein (see my 1988e).

'complete' symbols (see Linsky 1967a, 49–50), which should not need anything more to designate (like proper names), and this theory does not seem to be fulfilled by classes or by propositions.

Contextual elimination can be applied to classes only when, somewhat artificially, we add to them some predicate; exactly in the same way with descriptions, in whose case the requirement of to be part of a proposition is unavoidable (1910a, 173). But as for incomplete symbols, they have been previously defined as a class, as happens with numbers and other mathematical objects, but it is not easy to see why eliminative definitions must be 'in use' with them (apart of those corresponding to the respective classes). Concerning propositions, they are called now incomplete symbols, but not because of *being classes* (combinations of entities), but introducing the multiple theory of judgment and, *after that*, showing that the proposition is not a single entity, and calling incomplete symbol to the corresponding statement (here referred to as 'phrase') (1910a, 44).

Here Russell resorts to his rare doctrine that the statement can obtain a complete meaning by means of the assertion: 'the meaning is completed by the act of judging, and we no longer have an incomplete symbol' (1910a, 44). The Fregean sign of assertion (also a Pi) was destined to do this job, despite that this sign was also present in 1908a, while there the status of propositions did not suffer clearly this ontological attack. Finally, in 1908a, 84, assertion was applied to propositions and propositional functions (to their values), whereas in 1910a, 92, it was only applied to propositional functions.

We have, so, *three kinds* of incomplete symbols: (i) those that are to be defined in use (descriptions, classes and relations); (ii) those that must to be incomplete *qua classes* (numbers, series and other mathematical and physical objects), and require a previous definition in terms of classes, and (iii) propositions, which are transformed into genuine entities by means of our capacity of judging. Two consequences follow; firstly, it is not clear how constructive definitions 'analyze' a concept (for instance, the number) when we say, at the same time, that it is a 'mere' incomplete symbol. If, on the contrary, we admit the definition as only a lexical analysis, then it lacks philosophical interest. The truth is that the analysis should be placed (in the same example) in the point when we pass from the number to a class of classes, but in this way we cannot avoid the objection that we are merely handling synonymity between expressions which, *qua* incomplete symbols, are nothing, and we cannot analyze nothing (see my 1987b).

Secondly, the later introduction of physical objects as incomplete symbols (see my I988e) produces the strange impression that we must treat with care the words with which we designate them (for example, 'table'), perhaps looking for certain linguistic analysis destined to eliminate them (as suggested by Stebbing 1930a, 186). But physical objects can be incomplete symbols *only because they are classes*[28] (for example, classes of appearances). Thus their ontological status is like that of numbers. Anyway, the new theory came through once Russell, after Wittgenstein's criticisms of

[28] I find this thesis clearly formulated only by Smullyan (1958a, 241). Sainsbury (1979a, 239) almost formulates it, but Stebbing (1930a), Urmson (1956a) and Pears (1967a) do not see it.

1913 to his new theory of judgment, arrived at *neutral* logical constructions, namely to the rejection of his old belief that objects (mathematical or otherwise) are given to us by means of their structure. If we can construct several different structures, then the criterion for their admission as objects must be extra-logical.

All that coincided with the loss of the Platonic impulse produced by the discovery of problems in the principle of abstraction (only an old tradition after *Principles;* see my 1987b). Finally, Russell was led to drop the distinction between mathematical and philosophical definitions, mainly because incomplete symbols permitted only a kind of proper name very close to radical empiricism. But the general method of constructive definitions remained: that made possible the later assimilation of Whitehead's ideas (see my 1988c). The next passage, which seems earlier to the evolution of all this period, is a better sign of this continuity, for it was written in 1910 (1910b, 15).

> In this, as in all philosophical inquires, after a preliminary analysis of complex data we proceed again to build up complex things from their simpler constituents, starting from ideas which we understand though we cannot define them, and from premisses which, we know though we cannot prove them.

Acknowledgements I am very grateful to Ivor Grattan-Guinness for his kind help with my English and his valuable remarks and advice on the several versions of this article I have written. I am also grateful to Juan J. Acero, Gregory Landini, and an anonymous referee for comments on earlier versions. Although I cannot share the weight that Landini puts on the substitutional theory, his detailed and acute remarks have forced me to clarify my own interpretation and to improve my arguments. All unpublished material is contained in the Bertrand Russell Archives, McMcaster University, Hamilton, Ontario, Canada, who hold the copyright. My thanks are due to Kenneth Blackwell, the Russell Archivist, for the facilities and information provided.

References

The quotations in this article are referred always to the page number of the last reimpression (or translation) listed here.

Unpublished Writings by Russell

Correspondence with G. E. Moore. 1896–1905.
Correspondence with L. Couturat. 1897–1907 (only the few originals in McMaster, from Couturat's side. I have not seen the rest, in Switzerland).
m1904a 'On functions, classes and relations', 18 fols.
m1905a 'On substitution', 13 fols.
m1906a 'On substitution', 154 fols. (page numbers from 1 to 240).
m1906b 'The paradox of the liar', 125 fols. (page numbers from 1 to 108).

Published Writings by Russell

1901a 'Recent work on the principles of mathematics', *Int. month.*, 4, 83–101 = [as 'Mathematics and the metaphysicians'] *1918a*. 75–95.
1903a The principles or mathematics. Cambridge (University Press).
1904a 'Meinong's theory of complexes and assumptions', *Mind*, n.s., 13, 201–219; 336–354; 509–524 = *1973a*, 21–76.
1905a 'On denoting', *Mind*, n.s. 14, 479–493 = *1956a*, 41–56.
1906a 'On some difficulties in the theory of transfinite numbers and order types', *Proc. London Math. Soc.*, 4 (1), 29–53 = *1973a*, 135–164.
1906b 'On the substitutional theory of classes and relations', ms. published the first time in *1973a*, 165–189.
1906c 'Les paradoxes de la logique', *Rev. métaph. mor.*, 44, 627–650 [English original ('On "Insolubilia" and their solution by symbolic logic') in *1973a*, 190–214.]
1906d 'The theory of implication', *Amer. journ. math.*, 28, 159–202.
1908a 'Mathematical logic as based on the theory of types', *Amer, journ. math.*, 30, 222–262 = *1956a*, 59–102.
1910a (with A. N. Whitehead) *Principia mathematica*. vol. 1. Cambridge (University Press). [Reprinted 1925.]
1910b Philosophical essays. London (Longmans Green). [2nd. ed., London (Allen & Unwin), 1966.]
1911a 'Knowledge by acquaintance and knowledge by description'. *Proc. Arist. Soc.*, 11, 108–128 = *1918a*, 200–221.
1913a 'On the notion of cause', *Proc. Arist Soc.*, 13, 1–26 = *1918a*, 173–199.
1918a Mysticism and logic. London (Longmans Green). [Reprinted London (Unwin Paperbacks), 1976, with a different pagination.]
1944a 'My mental development', in Schilpp *1944a*, 3–20.
1956a Logic and knowledge (ed. R. C. Marsh), London (Allen & Unwin).
1959a My philosophical development, London (Allen & Unwin).
1973a Essays in analysis (ed. D. Lackey), London (Allen & Unwin).

Works by Other Authors

Bunn. R. *1980a* 'Developments in the foundations of mathematics, 1870–1910', in Grattan-Guinness *1980a*, 220–255.
Cavaillès, J *1962a Philosophie mathématique*, Paris (Hermann).
Chihara. C. S. *1972a* 'Russell's theory of types', in Pears *1972a*, 245–289.
Chihara. C. S. *1973a Ontology and the vicious circle principle*, Ithaca (Cornell University Press).
Clark. R. W. *1975a The life of Bertrand Russell*, London (Cape, and Weidenfeld and Nicolson). [Reprinted London (Penguin), 1978.]
Cocchiarella, N. *1980a* 'The development of the theory of logical types and the notion of a logical subject in Russell's early philosophy'. *Synthese*, 45, 71–115.
Coffa, J. A. *1979a* 'The humble origins of Russell's paradox', *Russell*, no. 33–34, 31–37.
Copi. 1. M. *1971a The theory of logical types*. London (Routledge & Kegan Paul).
Frege, G. *1976a Wissenschaftlicher Briefwechsel*, Hamburg (Meiner). [Partial English trans, by H. Kaal: *Philosophical and mathematical correspondence*, Oxford (Blackwell), 1980.]
Garciadiego, A. *1987a* 'Cantor and the paradoxes'. Unpublished paper kindly supplied by the author [whose Ph.D. thesis (*Bertrand Russell and the origin of the set theoretic paradoxes*. Toronto University), I have not yet seen. See also Moore].

Grattan-Guinness, I. *1972a* 'Bertrand Russell on his paradox and the multiplicative axiom', *J. phil. log.,* 1, 103–110.

Grattan-Guinness, I. *1974a* 'The Russell Archives: some new light on Russell's logicism', *Ann. sci.,* 31, 387–406.

Grattan-Guinness, I. *1977a Dear Russell—dear Jourdain,* London (Duckworth).

Grattan-Guinness, I. *1978a* 'How Russell discovered his paradox', *Hist. math.,* 5, 127–137.

Grattan-Guinness, I. *1980a* (ed.) *From the calculus to the set theory, 1630–1910,* London (Duckworth).

Grattan-Guinness, I. *1981a* 'Are there paradoxes of the set of all sets?', *Int.j. math. educ. sci. tech.,* 12, 9–18.

Grattan-Guinness, I. *1985a* 'Bertrand Russell's logical manuscripts, an apprehensive biief', *Hist. phil. log.,* 6, 53–74.

Hylton, P. *1980a* 'Russell's substitutional theory', *Synthèse,* 45, 1–31.

Kleene. S. C. *1952a Introduction to metamathematics,* Amsterdam (North Holland).

Kneale, W. and Kneale, M. *1962a The development of logic,* Oxford (Clarendon Press).

Landini, G. *1987a* 'Russell's substitutional theory of classes and relations', *Hist. phil. log.,* 8, 171–200.

Linsky, L. *1967a Referring,* London (Routledge & Kegan Paul).

Moore, G. H. and Garciadiego, A. *1981a* 'Burali-Forti's paradox: a reappraisal of its origins', *Hist. math.,* 8, 319–350.

Pears, D. *1967a Bertrand Russell and the British tradition in philosophy,* London (Collins).

Pears, D. *1972a* (ed.) *Bertrand Russell: A collection of critical essays,* New York (Doubleday).

Quine, W. V. *1966a* 'Russell's ontological development', *J. phil.* 63, 657–667.

Ramsey, F. P. *1931a The foundations of mathematics,* London (Routledge).

Rodríguez-Consuegra, F. A. *1987a* 'El método en la filosofía de Bertrand Russell. Un estudio sobre los orígenes de la filosofía analítica a través de la obra de Russell, sus manuscritos inéditos y los autores que más le influenciaron', Ph.D. thesis, University of Barcelona, x + 800 pp.

Rodríguez-Consuegra, F. A. *1987b* 'Russell's logicist definitions of numbers 1898–1913: chronology and significance', *Hist. phil. log.,* 8, 141–169.

Rodríguez-Consuegra, F. A. *1988a* 'Bertrand Russell 1898-1900: una filosofía de la matemática inédita', *Mathesis* (México), 4, 3–76.

Rodríguez-Consuegra, F. A. *1988b* 'Elementos logicistas en la obra de Peano y su escuela', forthcoming in *Mathesis,* 4. [pp. 221–299]

Rodríguez-Consuegra, F. A. *1988c* 'Bertrand Russell 1900–1913: los principios de la matemática, I', forthcoming in *Mathesis,* 4. [pp. 355–392]

Rodríguez-Consuegra, F. A. *1988d* 'Bertrand Russell 1900–1913: los principios de la matemática, II', forthcoming in *Mathesis,* 4. [pp. 489–521]

Rodríguez-Consuegra, F. A. *1988e* 'El impacto de Wittgenstein sobre Russell: últimos datos y visión global', forthcoming. [*Theoria* 7, nº 16–17–18, tomo B (1992): 875–911.]

Rodríguez-Consuegra, F. A. *1988f* 'El logicismo russelliano: origen, evolución y significado filosófico', forthcoming. [*Crítica* 67 (1991): 15–39.]

Rodríguez-Consuegra, F. A. *1988g* 'The origins of Russell's theory of descriptions according to the unpublished manuscripts', forthcoming in *Russell.* [*Russell: the Journal of the Bertrand Russell Archives* 9 (1989): 99–132.]

Rodríguez-Consuegra, F. A. *1988h* 'La influencia de Bradley en los orígenes de la filosofía analítica', forthcoming. ["Bradley y la filosofía análitica, I: significado, lógica y ontología". *Análisis filosófico* 11 (1991): 153–175. "Bradley y la filosofía análitica, II: teoría del conocimiento y metodología". *Análisis filosófico* 12 (1992): 51–78.]

Rodríguez-Consuegra, F. A. *1988i* 'Bertrand Russell 1895–1899: una filosofía prelogicista de la geometría', forthcoming in *Diálogos,* no. 55. [(1990): 71–123.]

Rodríguez-Consuegra, F. A. *1988J* 'La primera filosofía de Moore', forthcoming in *Agora.* ["La primera filosofía de Moore, I" (Teoría del juicio y ontología). *Ágora* 9 (1990): 153–170. "La primera filosofía de Moore, II" (Teoría del conocimiento y metodología). *Ágora* 10 (1991): 91–105.]

Sainsbury, R. M. *1979a Russell,* London (Routledge).

Schilpp, P. A. *1944a* (ed.) *The philosophy of Bertrand Russell,* La Salle (Open Court). [1971 (4th ed.).]

Smullyan. A. F. *1958a* 'Incomplete symbols'. *Phil. rev..* 67, 237–242.

Stebbing, L. S. *1930a A modern introduction to logic,* London (Methuen). [Spanish trans.: México (UNAM), 1965.]

Urmson, J. O. *1956a Philosophical analysis,* Oxford (University Press). [Reprinted 1969.]

Instituto S. Vilaseca, Spain

Chapter 2
The Evolution of Russell on Relations

Abstract My main goal here is to attempt a global account of Russell's views on relations from the logico-philosophical viewpoint, by considering these views as the very kernel of his fundamental problems and difficulties in trying to build up a consistent ontology. To do that I'll use Bradley's paradox against relations to check the different stages in Russell's evolution.

Keywords B. Russell · F.H. Bradley · Paradox · Relations · Logic · Philosophy

Mathematics Subject Classification Primary 0302 · Secondary 03A05

I shall maintain here that: (i) the attempt to regard relations (and similar notions) as terms was the main problem to solve in each stage of Russell's philosophical development; and (ii) Bradley's arguments against relations are useful to test Russell's several ways-out. Bradley's main argument, which I shall call "Bradley's paradox", was that relations are unintelligible as separated entities, for when we try to give an account of them, then we have to explain also the relation between them and the related terms, which give rise to an endless regress.

2.1 The First Views on Relations (1897–1900)

In *Foundations of Geometry* Russell claimed that geometry gave rise to an unavoidable "contradiction", for it needs, at the same time, absolute and relative notions, as shown by the fact that space can be regarded as a set of entities located in certain order, and also as the mere relational order involved. The way to solve this general contradiction was to introduce unextended "material" atoms providing the terms for spatial relations. In doing so, Russell was facing the difficulties of Bradley's doctrine about the ultimate non-reality of relations. That is why he regarded relations among spatial entities, in general, as "adjectives" of space, and that is why he arrived at the need to provide terms to these relations. Thus, at this stage, the problem can be

F. Rodríguez-Consuegra, *Russell, Gödel, Tarski*, Studies in Universal Logic,
https://doi.org/10.1007/978-3-032-08945-8_2

described as the opposition between relations in themselves and relations as adjectives of the related terms, but on further stages the same problem always underlay the new terminology and applications of Russell's philosophy.

The theory of relations to which Russell arrived in 1899 was completely different, it was the ground to overcome the contradictions of relativity, and it was introduced in "The classification of relations" (m1899), a manuscript which classifies these entities through two main properties, symmetry and transitivity. Apart from the classification in itself, Russell insisted on the irreducibility of asymmetrical relations to predicates of the related terms through this argument: in "A is before B", we have not predicates of A or B, for position cannot be a predicate, but nothing else apart from the point itself, which meant for him that relations are external.

This was the view later developed in other manuscripts, but the main problem concerning external relations was, admittedly, Bradley's paradox:

> I must confess that the above theory raises a very difficult question. When two terms have a relation, is the relation related to each? To answer affirmatively would lead at once to and endless regress; to answer negatively leaves it inexplicable how the relation can in any way belong to the terms. I am entirely unable to solve this difficulty, but am not convinced that it is insoluble… To solve this difficulty -if indeed it be soluble- would, I conceive, be the most valuable contribution which a modern philosopher could possibly make to philosophy.

This problem underlies Russell's philosophy from this moment onwards, and, as can be seen in the further stages, he never was able to provide a solution.

The main reason of this impossibility was that Moore's relational theory of judgment was already implicit in Russell's theory of relations. According to Moore, the proposition is a synthesis of two concepts related in some specific way according to their immutable nature. Of course this was supposed to involve, as a result, a "complex concept", but it is unclear where the unity of this complex lies. On the other hand, the pre-eminence of relations over the predicative pattern works merely by placing the two terms of the proposition on the same ontological level (i.e. by rejecting the duality between things and qualities), and this seems an strategy similar to Bradley's when he wrote that there are not independent propositions, S is P, but only propositions such that we can say "reality is such that S is P".

Besides, Moore's paradox of truth can be seen as a sign of his unsuccessful escape from Bradley's paradox. He wrote: "it is… impossible that truth should depend on a relation to existents or to an existent, since the proposition by which it is so defined must itself be true, and the truth of this can certainly not be established, without a vicious circle, by exhibiting its dependence of an existent" (1899a, 181). Thus, true propositions are simply those whose terms are combined with a "special" term: existence. The step Russell gave was precisely to regard relations as terms being able to directly combine concepts to one another, so that relations are second order terms, as existence was for Moore. But then the status of false propositions needs some explanation; are they mere nonexistent combinations of terms? Or, have them also some sort of "being" as a timeless part of the world? If the second, false propositions are to be admitted with the same status as true propositions; if the first, our minds create false combinations of existent terms. Russell chose the first solution, but it finally led him to a view of judgment as a multiple relation between the mind and the constituents of the object of judgment.

2.2 Principles of Mathematics (1901–1903)

Russell learnt from Peano enough devices to invent a logical calculus for relations. But he needed to preserve, for philosophical reasons, his "external" approach, and he thought that only the intensional account, contrary to Peano's (or Frege's), could do the task. Probably he believed that the extensional pattern, which interprets relations as classes of couples (i.e. as functions with two variables), was too close to the "monistic" view according to which the relation is a property of the related terms.

The explicit reasons were two. First, he says that the opposite view proceeds from the habit to regard relational propositions as more imperfect than predicative ones (§27). But this forgets: (i) that the Bradleyan and the mathematical approaches are two very different things; (ii) that Bradley ultimately rejected the predicative pattern, together with any relational approach, even an internalistic one. The second reason was the parallelism with the treatment of classes through propositional functions, which is rather intensional.

As for Bradley's paradox, Russell faces it by saying that the unity of propositions cannot be reconstructed through the analysis (§54), but he does not say where we have to locate this unity. For him it was necessary to decompose the proposition into relation (verb) and terms (subject and "predicate"). But this analysis makes it impossible to locate the claimed "formal" unity in any of the constituents: if we regard the verb as the special term containing this form, it seems that the rest of terms have no role; if we regard the verb as only a term, it seems that we need other relations to preserve the unity, with which we come to Bradley's paradox.

The strategy to escape is a division into two types of endless regresses: vicious and harmless (§§55, 99). The argument against the vicious one rests on saying that, although the need of more and more additional relations would transform relational propositions into nonsense, we have to distinguish between a "relating relation" and a "relation in itself", through what he calls "the indefinable element of assertion which distinguished a proposition from a concept" (§99). But this presupposes that we can separate form (the relating relation) and content (the relation in itself) by converting the former into one more element of the latter, and this involves in itself the same thing to be rejected, for the form preserves the unity of the proposition, while it is only one more constituent without any "additional" relation to the terms.

Thus, the surprising difference between the relating relation and the relation in itself was the ultimate recourse to face Bradley's paradox. Yet it is quite clear that the same difference can be made with the rest of what we may call "formal concepts" (following Wittgenstein), so that the difficulties with relations were only a particular aspect of a general problem: the one concerning what makes it possible for a unity to be something different from its constituents. Thus, we can speak not only of the relating relation and the relation in itself, but also of the class as one and as many, the predicate as concept and as term, the functional function, and the function as entity, and even of the proposition as complex unity and as mere aggregate.

Russell implicitly admitted the unity of the problem by pointing out that the contradiction of classes is only a form of a general contradiction, which affects to all the rest of relevant notions as well, i.e. relations, predicates, propositional functions and propositions. As we shall see, this can be used to show that Bradley's paradox, which points out precisely the impossibility of separating intension (the form, or relating relation) and extension (the content, or relation in itself) was involved, in some way, in all the rest of similar problems, as Russell later seemed to recognize by admitting Poincare's principle of the vicious circle.

2.3 From Principles to "On Denoting" (1903–1905)

At this stage Russell published writings show only one important theme: the epistemological consequences of the assimilation of Moore's theory of judgment, in the long article on Meinong. However, when we resort to the unpublished manuscripts we discover that there were two more branches of the same problem, through two long series of unpublished materials. First, a set of them about the meaning-denotation problem (more than 200 folios); second, an even longer set on the nature of functions (more than 500 folios). It can be shown that the three problems are different aspects of the general Bradleyan one about the relation between complexes and constituents.

2.3.1 The Nature of Functions

As for relations, Russell openly accepts now that ordered pairs capture their true essence ("Relations", m1903?, f. 5). This supposed the assimilation of relations to classes, in so far as logical resources are concerned. Yet Russell also points out the need for realizing that through the intensional symbol R, which is used for convenience, a different sense of relation as entity is indicated, "symbolic logic takes relations, like classes, in extension; and in this what are here called relations differ from what are commonly so called, in the same way in which classes differ from predicates" (*ibid,* f. 13). This is a sign of the need for a deeper philosophical treatment, which was made only in the manuscripts directly devoted to the nature of functions.

The manuscript "On functions" (1904 Oct.) contains an attempt to build up an ontology of complexes by introducing a new entity: the "mode of combination", it was destined to give an account of what gives a complex its unity: "a complex is determined by its constituents together with their mode of combination; it is not determined by the constituents alone". But the isolation of the new entity is impossible on pain of falling into Bradley's paradox (f. 1):

> The mode of combination of the constituents of a complex is not itself one of the constituents of the complex. For if it were, it would be combined with the other constituents to form the complex; hence we should need to specify the mode of combination of the constituents with their mode of combination.

Yet it seems this is only an ad hoc limitation, for Russell needs here three probably incompatible things. First, to give an account of the relational framework involved; that is why he adds that the mode of combination contains a *sense* or *order* as part of its essence. Second, to regard the mode of combination as an entity, because he is trying to reduce classes and relations to functions, so that functions must be something. Third, to explain how is possible to use functions of functions without contradiction, i.e. without allowing modes of combination be affected by themselves.

In trying to set up this last goal he wrote: "A mode of combination, like everything else, is an entity; but it is not one of the entities combined in the mode in question" (f. 2). This is another attempt of avoiding Bradley's paradox, but now in a way which permits us to see the parallelism with the contradiction of classes. For we can paraphrase the quotation as: "a class, like everything else, is an entity; but it is not one of the entities included in itself", by pointing out the differences between the class as one and as many; and the same thing can be done with other formal concepts.

In the manuscript "On the nature of functions" (ml904 Nov), the first question is again the old problem about the logical priority of relations and terms. On the one hand, for Russell complexes (terms) must be prior to functions (relations), for the function is the manner of combining the constituents, and "we cannot describe a way of combining constituents except by supposing constituents". But on the other hand, functions must be prior to complexes, for the general manner of combination (the function) is already a part of the description of any particular complex. He chooses the second view, for: "if we explicitly consider the mode of combination, it becomes itself a constituent of the complex, and is itself combined with the other constituents after a new mode of composition" (f. 2). But then he adds: "We are here in the region of the fundamental nature of complexity. The old problems of form and matter, of the possibility of analysis, of organic wholes., etc., are all forms of our present problem", which seems to suggest that there is no point in trying to solve the problem by adding further entities.

In "Plan for a general introduction to vol. II" (ml904, Plan) Russell says that his former arguments trying to extract a "formal" meaning from the usual notation for functions missed the point that the deep meaning of functions already underlies the variable itself (f. 813):

> the assumption of constant ϕ is already embodied in the use of an undefined x. So long as x remains undefined, ϕ"x *is* the constant function; ...Thus what results is that the postulate of functionality is embedded in the use of variables; hence it would seem to be presupposed in the notation, and therefore to be pre-formal.

The passage is important because it openly recognizes that the intension cannot be captured through any extensional device, but also because it points out that the meaning of functions is presupposed in any device we can use, so that formal logic cannot make it explicit. At any rate, the attempt finishes by recognizing that the difficult "something else" underlying functions cannot be reduced to further entities. And the reason is, again, Bradley's paradox, for in any case we should add some more way to indicate the positions of all our entities in the complex, which leads us to introduce additional relations among them (f. 848).

There is no clear end to these manuscripts, but it seems that Russell came to think that functions cannot be reduced to modes of combination, or any other complex entity: "in all the above, we have been starting from the complex and going to the function; but the opposite is the philosophically right course" (ml904 FN, 784). Thus, the recognition of functions as irreducible entities, which paved the way to their important role in the theory of types, is a sign that this first branch was closed by the impossibility of overcoming Bradley's paradox.

2.3.2 The Meaning-Denotation Problem

The relevant manuscripts ("On fundamentals") show several attempts to separate the psychological aspects of meaning by transforming it into an objective notion. But then it must be also an entity, so in some way it has to be a part of the proposition involved, as one of its constituents, in the same way that the entities denoted are already parts of the proposition as a complex. Thus we need some explanation about its relationship with the rest of constituents. That was the origin of the distinction between meaning-positions and entity-positions, which already poses a similar problem to that posed by functions.

However, the endless regress is already here, for when we try to speak about the meaning, we are putting it in an entity-position, so that we are really speaking of the denotation, not about the meaning. Thus, the key of the problem seems to be the mutual application of the two notions, so that the attempt to separate them forgets that intensions or forms cannot be regarded as *mere* extensions or terms. That is why Russell points out the impossibility of inventing a symbolism which avoids the need to distinguish the two sides in complexes.

Therefore, we have no longer one complex with the two aspects of meaning and denotation, but two entities, the complex, and its denotation (f. 39). Which seems to be the same as to admit that in a complex we have only the complex in itself, including the relations and other intensional entities which make it a complex, and the entities which are *parts* of the complex. Thus, it seems that Russell is here only recognizing that any attempt to forget this point is already condemned to failure. There are, of course, unclear points, as for instance that the denotation of the complex can be separated of the complex in itself. But I think this summary can be enough to show that some kind of Bradleyan argument was also involved in the meaning-denotation problem.

2.3.3 The Epistemology of Complexes

"Meinong's theory of complexes and assumptions" (1904) was Russell's first opportunity to delve deep into the underlying problems of Moore's theory of judgment through two main difficulties: to give an account of false propositions and also of the nature of judgment as a dual relation between mind and objects.

Bradley's endless regress was already involved in Meinong's claim that there is no presentation (before the mind) of two terms standing in certain relation, for this presupposes the relation to be already rightly related to the terms. Russell needs to deny this because he still believes that propositions are *objective* complexes, so that our minds can reproduce complexes with no thought of the relation in itself (f. 53):

> We can ... think of objects in a certain relation, without thinking of the relation at all. And apart from inspection, the endless regress seems to prove that this is so; for if not, we should have to think also of the relation of the relation to the terms, and so on.

It is unclear why Russell does not regard here the endless regress as "harmless". I think the reason is that he is starting to surrender to the view that in some sense it is only the mind which imposes relations to terms. A sign of this change is that he openly admits that in some difficult sense: "a relational proposition seems to be *about* its terms, in a way in which it is not *about* the relation", which led him to recognize that there is some true logical fact involved in the assertion that relations as less substantial in some sense than their terms (f. 53–54).

However, any "multiple-relation" theory is still discarded. Judgment cannot be a multiple relation of the mind with the several terms involved in the complex, because it would transform complexes into aggregates without unity. It seems, then, that Russell is resting on a new form of Bradley's paradox: if there were multiple relations of this kind, they never could reconstruct the original complex, presumably because we would need an additional judgment to state the correspondence. But then Russell has no account to give of false propositions, and he escapes towards the comfortable intuition: we merely apprehend truth and falsehood, which are incapable of analysis.

"On denoting" affected to the three branches at the same time, as for the nature of functions as forms, it provided the distinction between real and apparent constituents of complexes, together with a device to eliminate the "form" of certain constituents, clarifying its real role in other complexes. Concerning the meaning-denotation problem, it pointed out the difficulties of the distinction in itself, which meant that we cannot separate the intensional and the extensional aspects of complexes. However, as descriptive complexes are significant when they occur as constituents of other complexes, we still need to give an account of propositions. As for our knowledge of true and false complexes, there was an implicit change: now we can constitute "false complexes" by combining mere concepts in certain ways, which is a clear precedent for the "multiple-theory".

2.4 The New Theory of Judgment (1906–1912)

In this stage Russell was trying to draw the consequences of the new theory of descriptions in order to avoid the contradictions. A first attempt was to return to the substitutional ideas, by avoiding classes and functions in favor of propositions and constituents. But the impossibility of dispensing with propositional functions, and the difficulty of giving an account of the nature of propositions, led Russell to the abandonment of the substitutional viewpoint, and even to dispense with propositions as incomplete symbols. This, again, involved the need for a new theory of judgment, which was already *accepted* and developed in manuscripts of 1906–1907. However, the new theory involved similar problems.

2.4.1 Propositions and Statements

To solve the paradox of the Liar, Russell introduced the distinction between propositions and statements, i.e. between propositions without and with apparent (bound) variables. Thus, the Liar means: "there is a proposition *p* which I am affirming and which is false", and this is only a statement, not a proposition. But then our problems already begin, for in Bradleyan terms we cannot set up a property of a complex because this involves another complex embracing the first and trying to relate it and its constituents, and so on.

I think Russell's device is meant only to distinguish between acceptable and unacceptable complexes. And this is not very useful, for we still need a further prohibition concerning statements as not being able, again, to be true objects. That is why Russell immediately adds that we cannot say "I am making a statement which is false": we cannot make general statements "without specifying a particular number of apparent variables", and this must be done through some hierarchy. Yet this causes similar problems, for the admission of statements was made only in order to allow "general" propositions, no matter how different the new name is. And we still need a prohibition in some point, as here in between statements and general statements. Accordingly, there must be a deep problem hidden in the very nature of propositions. That is why this is one of the main concerns in the relevant manuscripts of 1906.

Ultimately, these problems led Russell to another theory of judgment, where there are constituents, but no longer propositions as true unities. This was made in the ms "The paradox of the Liar" (ml906b), which definitely stated the need for propositional functions, and provided the new theory to explain how is possible that propositions can be the values of propositional functions. Yet the recognition that false propositions are a product of the mind was already the seed of the following step, that all propositions are nothing but aggregates without unity. Thus, when Russell wants to escape to one form of Bradley's paradox, he falls into another.

2.4.2 Other Manuscripts from 1906 to 1907

In a long set of "unclassified notes" (ml1906? Uncl.), the problem is again faced in the context of the multiple theory: "when we deny what can be truly denied, the proposition denied does not enter in, but is broken up and psychologized" (f. 154). This cannot be successful unless we extend this result to all propositions, which appears in a notation that is the first anticipation of the one used for the multiple theory seven years later ($\mathfrak{J}$ (A, α, x), f. 174), for it includes symbols for: the "multiple relation", the mind, the form of the proposition involved, and the argument to which some property is applied according to that form. Thus, it is admitted that propositions *in general* can be explained by means of constituents, and therefore that although we may have complex knowledge, we never have knowledge of complexes (f. 142).

But the introduction of complexes as mere complexes of ideas, although it can be a solution to the old problems, can be regarded as a clear escape to Bradley's paradox, precisely in the only way Bradley himself already stated: by recognizing that the analysis of complexes is only a mental task. However, the new attempt falls, again, into Bradley's paradox, for we cannot, according to the old master, consider complexes on the same ontological level than constituents, and it is obvious that Russell is thinking of such a solution, which can be seen in the notation he uses, where the "form" of the complex is one of the explicit constituents.

However, the line is definitely stated in the manuscript "Fundamentals" (ml907 Jan), which shows that the first theory of types was published in 1908 with an underlying rejection of propositions as entities: "A belief has several objects; when correct, it has, in a secondary sense, the complex as its object. *Belief* means a certain kind of relation to several objects" (ff. 4–5). Besides, the definite notation for the multiple theory is clearly stated:

$$f(\text{A}, \text{x}, \phi) . = . \text{ A believes x has the property } \phi.$$

However, we find no attempt to face the philosophical problem of justifying the new theory, as involving a new challenge against Bradley's paradox.

2.4.3 The Published Views

In the first publication about this theme, from 1907, Russell regarded the new theory as only one possibility, but it was officially accepted in 1910. However, although propositions are now incomplete symbols, and need the judgment to hove meaning, it is unclear whether it means that propositions *are* only a mere creation of the mind.

The problem lies in explaining how is possible that there is no complex which is the object of judgment, while some theory of truth as correspondence is maintained, and the subsequent paradox is avoided. For Russell there is no true complex "ArB" in itself, but "I believe that ArB", so that the possibility of some correspondence is already obscure. The judging relation imposes its sense and then produces actually a complex, which does not exist in itself, so that it is hard to see that there must be a correspondence between the judging relation and the secondary complex, which are both a product of the mind. The endless regress is present anyway: to explain the relation of correspondence between the first complex (the judging relation), and the second complex (the secondary relation), we need another belief stating the fact of the correspondence.

The climax of all these problems can be found in the articles where Russell considers the status of relations and predicates, as for instance in "The basis of realism" (1911c). Russell starts by considering the difference between monadic and dyadic concepts, doubtless to face the difficulty, but then he is forced to speak of predicates as having the two different natures, "it is of course the case that, whenever a subject has a predicate, there is a dyadic relation of subject and predicate, but it does not follow that there is not also a proposition in which the predicate is merely predicated" (f. 159). Thus, the analogy with relations is complete: we have predicates as relations and predicates as terms, which leads us to exactly the same problems. The first form avoids Bradley's paradox, but cannot provide terms; the second form provides true terms, but cannot explain complexes, unless we introduce further relating constituents and violate Bradley's paradox.

Russell made a further effort to maintain the distinction by adding: "Whenever *a* has the relation *R* to *b*, there is a triadic relation of *a* and *R* and *b*, but in this relation *R* occurs as a term of the relation, not as the relating relation of the proposition". But the new manoeuvre is only the old strategy, for then there are two additional relations which do not appear in the notation: the triadic relation actually relating the rest of the constituents; and the relation R no longer as relation but as term. Thus, to explain a dyadic relation we need a triadic one where the former acquires a different status, so that presumably to explain the triadic relation we are going to need a quadruple one, and so on.

2.4.4 The Polemic with Bradley

The published polemic was started by Bradley in 1910, when he accused Russell of maintaining an inconsistent pluralism, in admitting complex unities which cannot be analyzed into terms and relations. Russell's response tried to avoid the ultimate inconsistency by saying that he did not maintain unities as incapable of analysis, but only that the mere enumeration of the constituents cannot reconstruct the unity. "A complex differs from the mere aggregate of its constituents, since it is one, not many, and the relation which is one of its constituents enters into it as an actually relating relation, and not merely as one member of an aggregate". But then he provides no solution to the objection, as Bradley's reply lucidly points out:

> Is there anything, I ask, in a unity beside its "constituents", i.e. the terms and the relation, and, if there is anything more, in what does this "more" consist? Mr. Russell tells us that we have got merely an enumeration or merely an aggregate. Even with merely so much I

> should still have to ask how even so much is possible. But, since we seem to have something beyond either, the puzzle grows worse.

The personal correspondence shows Russell as much more appreciative of Bradley's philosophy. In 1907 Bradley already pointed out that that wholes can by no means be reconstructed, for they are nonrelational in the last analysis, "if you break this entity up, and set down any part as independent, then, starting with this part, there is no getting beyond it except arbitrarily… You will say that you replace this by external relations. But it is denied that these serve" (Oct. 21, 1907). Russell admitted the difficulties, but only by claiming that relatedness does not imply complexity, with no further explanation of how can we give an account of relations as *related* to their terms (Oct. 29, 1907).

The final stage of the correspondence is very well shown in this passage by Russell, which contains the failed attempt to escape the old paradox: "I do not consider pluralism incompatible with the existence of complex entities, I consider that in every case where two simples have a relation, there is a complex entity consisting of the two simples so related" (April 9, 1910). But this, again, misses the point, for the expression "complex entity" involves different unexplained senses. We have only to replace here "simples" by "terms" to realize that the relation is also a term on Russell's own view, so that he has to explain the difference between the complex as constituted by the three terms, and the complex as being the "complex entity", supposedly formed by the terms as "simples" and the "relating relation". That is why Russell finally added a third entity, the "form", and faced the same paradox at a upper level.

Russell's final attitude implied the recognition that the problem of unities was really unsolved (March 2, 1911). Fortunately there is a passage showing the strong connection among: Bradley's criticisms, Russell's unsolved difficulties, the new theory of judgment and Wittgenstein's objections, which is the only place where Russell admitted these important links (Jan. 30, 1914):

> I fully recognise the vital importance of the questions you raise, particularly as regards "unities"; I recognise that it is my duty to answer if I can, and, if I cannot, to look for an answer as long as I live... Chiefly through the work of an Austrian pupil of mine, I seem now to see answers about unities; but the subject is so difficult and fundamental that I still hesitate.

2.5 The "Multiple" Theory Staggering (1913–1918)

This stage can be regarded, to some extent, as the one where Russell's philosophy had to pay the price for the continuous delay in facing the main problem. And this price was expensive, for it involved the abandonment of a major project in philosophy of logic, and then of the hope of finding an acceptable global philosophy. But the difficulties which made the project impossible were already present in former stages, despite the fact that Wittgenstein's criticisms were the apparent main cause of Russell's disappointment with his own philosophy.

2.5.1 The Unpublished Epistemology of Logic

The manuscript "What is logic?" (ml912) is already an attempt to characterize the notion of form. But the starting point returns to former views: "The *form* of a complex is what it has in common with a complex obtained by replacing each constituent of the complex by something different". Thus, we have to avoid Bradley's paradox, for if we make the form a constituent: "it would have to be somehow related to the other constituents, and the way in which it was related would really be the form; hence an endless regress" (f. 2). Therefore, although no final definition of the notion is provided, Bradley's paradox is respected, and for the same reason any possible violation of the theory of types was apparently avoided, for any attempt to put the form among the constituents would be such a violation.

That is why it is so strange that in Theory of knowledge (1913), Russell's main recourse was the admission of forms as constituents of complexes with no explanation of the type-theoretical problems involved. Of course this was also an attempt to avoid the former criticisms by Stout, Bradley, and Wittgenstein, which forced him to introduce some changes to avoid the rather idealist consequences of regarding complexes as a mental creation. But this criticisms forced Russell to face a very unpleasant dilemma: either maintaining the old theory of types by abandoning the multiple theory of judgment and by renouncing the attempt to characterize complexes, and then logic, or abandoning some philosophical consequences of the theory of types by complementing the multiple theory with an explicit device making the admission of forms as some sort of constituent possible. Russell's solution in 1913 was clearly a compromise between the two horns of the dilemma, but it seems to involve rather the second alternative, despite Bradley's prohibition.

Russell now rejects the view that the form can be a "mere" constituent of complexes with the usual Bradleyan argument. Thus, in "Socrates is human", "is" represents the form, and then cannot be a constituent: "for if it were, there would have to be a new way in which it and the two other constituents are put together, and if we take this way as again a constituent, we find ourselves embarked on an endless regress" (p. 16). But when he tries to explain the nature of form, he rejects not only to regard it as an equivalence relation (i.e. "to have the same form"), but also as a mere primitive idea, for it would lead us to the usual paradox, when trying to relate this idea with the others in the system. The solution was to regard it as an indefinable object corresponding to certain expressions.

Thus, the form of subject-predicate complexes will be "something has some predicate", and the form of dyadic complexes will be "something has some relation to something". Russell tries to avoid the attack of circularity by writing: "in spite of the difficulties of language, it seems not paradoxical to say that, in order to understand a proposition which states that x has the relation R to y, we must understand what is meant by "something having some relation to something" (p. 114). However, the compromise supposed only one change in the general pattern of the already *published* multiple theory of judgment: to incorporate a symbol for the form to the general complex constituting the judgment (e.g. "U (S, x, R, y, γ)"), returning thus to the view of the manuscripts from 1906 to 1907. However, this is not only an attempt to leap over Bradley's paradox, but also to give an objective status to the form, as something with which we are acquainted. The important change was to show it as absorbing the "relating" relation, but we still need to explain the status of the form into the complex.

I cannot enter here in Wittgenstein's criticisms in themselves, but for him the main defect of Russell's revised theory was that all the difficulties involved have their common root in the attempt of regarding the form as a new constituent. Wittgenstein expressed this in a language infected with mystical connotations, but this cannot avoid the fact that he clearly saw the difficulties of Russell's attempt of trying to fix the nature of a form without using, at the same time, another form. Thus, Bradley's paradox is also present: we cannot *speak* with sense about the relation between the judging complex and the secondary complex.

2.5.2 The Published Views on Forms

At this stage Russell, rather surprisingly, abandoned neither some version of the multiple theory, nor the regarding of forms as constituents. In *Our knowledge of the external world,* Russell starts by declaring forms as the object of "philosophical logic" (p. 52)., although he says only that "form is not another constituent, but is the way the constituents are put together". Thus, with no word about his failure in constructing an epistemology of logic, he even adds the claim that we have some kind of knowledge of logical forms, "which allows us to understand sentences" (p. 53). We have thus the usual arguments, but nothing about the status of forms in the judging complex, while some sort of multiple theory seems to be maintained, for when Russell denies negative facts except as false beliefs, he writes: "it is therefore necessary, in analysing a belief, to look for some other logical form than a two-term relation" (p. 66).

"The philosophy of logical atomism" (1918) was Russell's last attempt to maintain a consistent view before officially abandoning the multiple theory. As regards forms, the same idea of constituting an inventory is introduced, by only changing "forms of propositions" by "forms of facts" (p. 216). This is presented as a "realistic bias", despite the need for providing an account of false "facts" in terms of some kind of multiple relation. And the multiple theory seems to be *still maintained here,* for it is said that belief is not a dual relation (p. 224): "Therefore the belief does not really contain a proposition as a constituent but only contains the constituents of the proposition as constituents". Bradley's paradox is present when Russell admits that we cannot put the verb on the same level than its terms because it is an instance of form, i.e.: it "is not a constituent of the proposition. If it were you would have to have that constituent related to the other constituents". Thus, Bradley's paradox makes the form as a constituent impossible, while this is needed in the multiple theory.

Finally these ideas also appear in *Introduction to Mathematical Philosophy* (pp. 198 ff), written in 1918. Here Russell needed some definition of logic, for the book was explicitly devoted to the logicist foundations of mathematics, so that he resorts to the usual "solution": "we may accept, as a first approximation, the view that *forms* are what enter into logical propositions as their constituents", so that "Logic is concerned only with *forms,* and is concerned with them only in the way of stating that they are always or sometimes true". Thus, he says not only that forms are constituents, but also that we can have higher propositions in which we refer to these forms and the rest of the constituents. Therefore, he is forgetting the two forms of Bradley's paradox, despite the fact that, some lines before, he admitted the impossibility of regarding the form in itself as a new constituent, on pain of the usual endless regress.

If I have to propose some conclusion, I can add that the rather chaotic philosophy usually called "logical atomism", can be hardly regarded as "logical" or as "atomism", it was rather the compromise between two *ontological* incompatible traditions: that from Moore, and that from Bradley. And it was rather the attempt to include relations in the ultimate inventory of the world, which was impossible as it was attempted by making them to have a twofold nature, that of terms, and that of non-terms. Furthermore, this philosophy was hardly realistic, for the idealistic implications of the multiple theory of judgment made it impossible to maintain that complexes are genuine realities. They were, at most, the result of *our* application of forms to the terms we perceive. But as these terms can be further constructed out of abstract structures, which ultimately are nothing but "incomplete symbols", then the old realistic world is replaced by an idealistic and holistic one, where terms can no longer be opposed to relations, relations can no longer be regarded as true terms, and the seed of the later abandonment of the distinction between subject and object was already present. The next, and last, stage was only the drawing of the implicit consequences.

References

Since the manuscript of this paper was the text of a lecture, the bibliographical references were minimum. In the following, I try to reconstruct them by resorting to the full bibliography of the book which was developed by starting from this text, which was published much later: *Relational Ontology and Analytic Philosophy: Bertrand Russell and Bradley's ghost.* Details of the book can be found here: https://www.amazon.com/dp/1709974710.

All the unpublished manuscripts mentioned here have been published in the massive series *The Collected Papers of Bertrand Russell*, still in progress. All details about that magnificent series can be found here: https://russell.humanities.mcmaster.ca/our-work/collected-papers/. All the items I mention in the text can be easily found in the corresponding volumes, by using title, year of composition and number of folios.

Russell's Unpublished Manuscripts

"The classification of relations" (m1899): *m1899b* "The classification of relations", 16 f., ts.

"Relations", ml903?: *m1903b* "Relations", 13 f.
"On functions" (1904 Oct.): *m1904d* "On functions", 30 f.
"On the nature of functions" (ml904 Nov): *m1904f.* "On the nature of functions", 16 f.
"Plan for a general introduction to vol. II" (ml904, Plan): *m1904h* "Plan for a general introduction to vol. II", 72 f.
m1904 FN, 784: *m1904i* "FN", 233 f.
"On fundamentals": *m1905d* "On fundamentals", 81 f.
"The paradox of the Liar" (ml906b): *m1906b.* "The paradox of the liar", 125 fols.
"unclassified notes" (ml906? Uncl.): *m1906d* "[unclassified notes]", 110 f.
"Fundamentals" (ml907 Jan): *m1907a* "Fundamentals", 68 f.
"What is logic?" (ml912): *m1912* "What is logic?", 5 f.

Russell's Published Works

Foundations of Geometry 1897a An essay on the foundations of geometry, Cambridge: University Press.
Principles of mathematics 1903 The principles of mathematics, Cambridge (Univ. Press).
"Meinong's theory of complexes and assumptions" (1904) *1904* "Meinong's theory of complexes and assumptions".
Mind 18: 204–19; 336–54; 509–24.
"On denoting" *1905* "On denoting". *Mind* 14: 479–93.
"The basis of realism" (1911c) *1911b* "The basis of realism". *Jrn. Philos.,* 8: 158–61.
Theory of knowledge (1913), *1984 Theory of knowledge: the 1913 manuscript,* ed. E.R. Eames and K. Blackwell, London: Allen & Unwin.
Our knowledge of the external world 1914 Our knowledge of the external world, London: Allen & Unwin
"The philosophy of logical atomism" (1918) *1918b* "The philosophy of logical atomism". *The Monist,* 28–29 (1918–1919). Rep. in *1956. (1986:* 157–244.)
Introduction to Mathematical Philosophy 1919a Introduction to mathematical philosophy, London: Allen & Unwin.

Works by Other Authors

Bradley, F.H. *1910* "On appearance, error and contradiction", *Mind* 19: 154–185.
Moore, G.E. *1899a* "The Nature of Judgment". *Mind* 8, 176–193.

Correspondence

Russell-Bradley correspondence 1907–14. In the Russell Archives 1900–1922.

Chapter 3
Russell's Unwritten Reply to Gödel

Abstract During many years there has been controversy about Russell's famous note at the end of his "Reply to criticisms" in the Schilpp volume (*The Philosophy of Bertrand Rusell*), dated on July 1944, according to which he was unable to reply to Gödel's contribution to the volume. I took advantage of my research stay in the Russell Archives to try to follow the track of Russell's assertion looking for historical evidence.

Keywords B. Russell · K. Gödel · Logic · Mathematics

Mathematics Subject Classification Primary 01A70 · Secondary 00A30

Like many other people, I was always very curious about the facts underlying Russell's brief explanation, as well as about the apparent lack of a relationship between Russell and Gödel. The matter is not philosophically important it itself, but is related to Russell's personal honesty. Here and there some suspicions (e.g. Hao Wang, *Reflections on Kurt Gödel*, 1987) have been spread that Russell might have already seen the first version of Gödel's article (which was written in May). If so, it is openly suggested that Russell had enough time to reply in due course, so that his explanation was only an excuse. In the Russell Archives I found quite a lot of information related to our problem.

I found first a draft by Gödel, which contributes to understand the lack of Russell's reply. The letter, whose original is now in Gödel's *Nachlass* in Princeton, and to my knowledge it has never been published,[1] is dated on Sept. 28, 1943, and it says:

> Dear Prof. Russell, I regret very much that (as Prof. Schilpp has informed me) you don't intend to write a reply to my article in the Library of Living Philosophers, and am very sorry that this is apparently due to my sending in the manuscript so late. On the other hand I still hope that you may perhaps change your mind, since your decision seems to have been based on the wrong assumption that my article will not be controversial. This is

[1] That was true when I wrote the text of this lecture. Finally all the Gödel correspondence was published much later, in the last two vols. of the magnificent series Kurt Gödel: *Collected Works*, OUP.

F. Rodríguez-Consuegra, *Russell, Gödel, Tarski*, Studies in Universal Logic,
https://doi.org/10.1007/978-3-032-08945-8_3

by no means so. I am advocating in some respects the exact opposite of the development inaugurated by Wittgenstein and therefore suspect that many passages will contradict directly your present opinion. Furthermore I am criticizing the vicious circle principle and appendix B of Principia, which I believe contains formal mistakes that make the proof invalid. The reader would probably find it very strange if there is no reply, not to mention that the reader of this volume will naturally be more interested in your opinions than in mine, in fact I doubt very much if my article should be published at all without a reply at least to the main items [there is here a deleted line: "Frankly speaking I am in favor of not publishing..."] and you will perhaps agree with me after having read it.

Very sincerely yours. Kurt Gödel.

P.S. I have sent in the manuscript today. I did not send it directly to you because it is not very easily readable in its present condition.

This letter already proves that Russell did not know by September the article, but it is unclear about the facts concerning the first version of the article, and concerning Russell's supposed opinion that the article was not "controversial".

Hao Wang also says that Gödel wrote Russell to convince him to write some reply, so that, according to him, the mentioned letter to Russell must have been actually sent, although the original of the actual letter is missing in the Russell Archives. Besides, he tells us that Gödel did not finally withdraw his paper because he "eventually yielded to Schilpp's entreaties". On the other hand, he adds that it is strange that, given that Russell's original Reply to the criticisms was dated on July 1943, he was unable to write a reply to Gödel's first version, which was finished and sent to Schilpp in May 1943, and then he writes, "it is not clear that Schilpp did not send him the May version". Hao Wang finally includes the following remark, rather derogatory for Russell: "At any rate, it appears that Russell was at that time probably no longer capable of dealing with G's paper judiciously" (p. 111).

I looked for some more evidence in the correspondence between Russell and Schilpp in 1943, and I found quite a lot of new information. On June 26 Schilpp says he is working on Gödel's first version, and that he will send it to Russell as soon as he has Gödel's approval. On June 28 he encourages Russell to start to write the reply despite the absence of Gödel's essay, and the same in July 3, where he adds that the essay was "quite technical". With that we have:

(i) a further proof that Russell did not see the May version, and
(ii) that Schilpp's description must have been Russell's reason to think that the article was not "controversial", although the version we now know is rather philosophical than technical (in the logical sense).

On July 5 Schilpp says he has already returned the first version to Gödel, and conjectures that perhaps he will send the final revision to Russell about 10 days later. On July 13 Russell says he has nearly finished a 20.000 words Reply, and on July 16 Schilpp writes: "Unfortunately he [Gödel] is taking much more time than I expected in revising his essay, and seems to be unwilling to have us let you see his essay in its present form", and then he conjectures a new period of a few days. On July 29 Russell writes that his Reply would be increased by the part devoted to Gödel. Finally, on July 28 Schilpp says he sent a "S.O.S. letter" to Gödel, although he adds: "Gödel does not seem to be the sort of person to respond to any S.O.S.'s or any other kind of pressure. So: all we can do is wait". With that we know that Gödel's delay, together with his reluctance to show the first version to Russell, were really the reasons of the problem, for it is clear that Russell was decided to include a reply to him.

However, Russell must have been tired of waiting, so that he decided to send the Reply. On Aug. 5 Schilpp acknowledges receipt and says he sent Gödel a telegram, to which he replied: "Manuscript ready in some days unless delayed by army service examination tomorrow. Gödel". Schilpp adds: "I do not know what "some days" means; it has been already been "some weeks" since I sent him the revised copy of his original MS... I keep my fingers crossed". Russell's next letter already contains the decision not to reply to Gödel, as well as the conjecture about his content: "I think I will say nothing about Gödel except a P.S. that it came too late. In any case it is not likely to be controversial".

Schilpp's reply was limited to recall Russell, on Aug. 11, that as Gödel was "a refugee from oppression", surely his lateness was mostly due to his problems with the English. He finally adds: "I am sure he would greatly appreciate something from you by way of reply, however pointed and brief it might be". But Russell insists, on Aug. 14, in the previous idea of a final note: "As for Gödel, you could insert a note saying his essay came too late". With that, we have already the proof that Russell decision was clearly motivated by Gödel's incredible delay (he had only to revise the article), and once more that this decision was taken before he knew the article. (By the way, it is interesting to know that Gödel caused similar problems with other contributions to Schilpp's collection: his paper was the last one to reach the editor for the Einstein volume, and the one intended for Carnap's was never sent, after endless revisions.)

As we know through Gödel's letter to Russell, he finally sent the revised version on September 28. Schilpp immediately sent it to Russell according to Gödel's "specifically expressed desire", and wrote: "Should you desire to reply to it -however briefly or at whatever length- I am sure that this can still be inserted at the proper place in your formal "Reply";... But I shall leave this entirely up to your best judgment and decision" (Oct. 2). Russell's answer of Oct. 5 contains the last element of the story.

> I have received Gödel's article, and a letter from him urging me to answer it. It is quite impossible for me to make a detailed answer. I have not worked at mathematical logic since 1927 and it would take me at least a month's work. I am prepared to write a short paragraph saying I am unable to form a critical estimate of his article, but I think it quite probable that most of his criticisms are justified. I hope this will satisfy him and you.

Thus, we know now (i) that Russell did receive Gödel's letter, (ii) that he persisted in his former decision not to write a reply, after four months waiting for being able to see some version of Gödel's paper, and (iii) that in such circumstances he did not want to spend one more month in preparing an appropriate reply. The rest of the correspondence shows Schilpp's approval of Russell's decision (Oct. 7), and Russell's preference to locate the note at the end of his Reply, although through this last letter we also know that Russell never answered Gödel's letter: "Please let him know, as he wrote to me, and I have no time to write to him, being desperately busy" (Oct. 11).

The only later contact between Russell and Gödel took place in the Spring of 1944, when they met in Einstein's home at Princeton. It can be conjectured that this contact could hardly be very fruitful, after the first and not very fortunate chapter of the story. But there is also some more information about a second and final chapter, it took place when Ken Blackwell sent Gödel a letter on Sept. 22, 1971, asking about some possible correspondence with Russell, as well as for some opinion about Russell's passage in his *Autobiography,* where Russell tells us about his discussions in Princeton with Einstein, Pauli and Gödel. In the passage Gödel is mentioned as an "unadulterated Platonist", for defending the true existence of the logical constants (e.g. of the "not"). But, rather unfortunately, he is also mentioned as a Jew, together with Einstein and Pauli, and all of them are told to have "had a German bias toward metaphysics", in spite of the fact that all of them were "in intention, cosmopolitans".

Gödel never answered the letter, but in the Russell Archives there is a draft from Gödel's *Nachlass* of a never sent reply to Blackwell. Hao Wang did publish partially this draft on his book on Gödel in 1987, after Dawson did it in 1984, and Feferman included a brief reference to the matter in his introduction to Gödel's *Collected Works.* The complete reconstructed text says:

> Replying to your letter of [September 22] I would like to say that I never had any correspondence with Bertr[and] R[ussell]. As far as the passage about me is concerned, I have to say *first* (for the sake of truth) that I am not a Jew (even though I don't think this question is of any importance), 2) that the passage gives the wrong impression that I had many discussions with Russell, which was by no means the case (I remember only one). 3) Concerning my "unadulterated" Platonism, it is no more "unadulterated" than Russell's own in 1921 when in the *Introduction* [*to Mathematical Philosophy,* first published in 1919, p. 169] he said "[Logic is concerned with the real world just as truly as zoology, though with its more abstract and general features]". At that time evidently Russell had met the "not" even in this world, but later on under the influence of Wittgenstein he chose to overlook it.

Leaving aside the fact that Russell was no longer a Platonist in 1918, when the book was actually written, at least in the same sense he was in *Principles,* the fact is that the tone of the draft seems somewhat uncomfortable and, as a whole, Gödel seems to me rather annoyed with the whole matter. To my knowledge, there is no additional evidence explaining the lack of a fruitful relationship between Russell and Gödel, probably the two greatest logicians of this century.

Chapter 4
Russell, Gödel and Logicism

Abstract In this paper I discuss three assertions, which are sometimes taken for granted: 1. Russell did not understand Gödel's celebrated results, which he interpreted as implying that arithmetic is inconsistent, and also as new "puzzles" to be solved through the theory of types; 2. Russell was unable to handle the distinction between theory and metatheory, and 3. Russell's logicism collapsed after Gödel's incompleteness results. By taking into consideration relevant—published and unpublished—materials, I will show that the three assertions are either false or considerably open to doubt (Due to limitations of space in these Proceedings, I was forced to delete the first section of the original paper, as it was actually read. That section dealt with the reasons why Russell never published a reply to Gödel's contribution to Schilpp, P. 1944 (ed.) The philosophy o/ Bertrand Russell. La Salle, Ill.: Open Court, 1971. [That section now recovered as Chap. 3 of this volume].).

Keywords B. Russell · K. Gödel · Mathematics · Logic · Logicism

Mathematics Subject Classification Primary 03A05 · Secondary 00A30

4.1 I

Russell did not understand Gödel's celebrated results, which he interpreted (i) *to imply that arithmetic* is *inconsistent, and also* (ii) *as new "puzzles" to be solved through the theory of types.*

The supposed basis for (i) is a letter from Russell to L. Henkin of April 1, 1963 (now in the Russell Archives) in which he wrote "I realized, of course, that Gödel's work is of fundamental importance, but I was puzzled by it. It made me glad that I was no longer working at mathematical logic. If a given set of axioms leads to a contradiction, it is clear that at least one of the axioms must be false. Does this apply to school-boy's arithmetic, and, if so, can we believe anything that we were taught in youth? Are we to think that 2 + 2 is not 4, but 4.001? Obviously, this is not what is intended." Also, Russell asked Henkin to tell him about the way in which

F. Rodríguez-Consuegra, *Russell, Gödel, Tarski*, Studies in Universal Logic,
https://doi.org/10.1007/978-3-032-08945-8_4

ordinary mathematics is affected by Gödel's work. In his response (July 17; also in the R. A.), Henkin wrote: "...Gödel did *not* show (what seems to be suggested by your first paragraph) that the systems he investigated were inconsistent. Rather he showed these systems to be incomplete...". As for ordinary mathematics, he added: "every working mathematician has heard of Gödel's work, but (except in the field of logic) there have been no discernible consequences on mathematical practice". As a whole this interchange has been interpreted by Dawson 1984 to show Russell as believing that Gödel's results imply that arithmetic leads to contradiction. However, Dawson told me in a personal communication (February 23, 1990) that more charitable interpretations are possible. Let me then attempt one of these charitable interpretations.

For one thing it is of course possible that Russell misinterpreted Gödel's incompleteness theorem and also the unprovability of consistency. At any rate he would not be the only one in so doing, especially after a first impression (see Ladriere 1957 and Dawson 1984 for details about many more misinterpretations). But if we suppose that for Russell Gödel proved arithmetic to lead to contradictions, I think we are forced to provide some line of argument leading from Gödel's incompleteness theorem to some form of particular contradiction. The most similar line I have found in the literature probably known to Russell is Black's account of Gödel's results in his 1933 book. Black wrote that, in his (1931) article, Gödel showed that a proposition, which we will call *G*, although obviously true, could not be formally demonstrated, and that any demonstration of *G* "would lead to a contradiction" (p. 167). This could be interpreted as saying that the truth of *G* leads to a contradiction, for it implies that its provability is equivalent to the provability of its negation. But this is not so. If we suppose arithmetic to be consistent then neither *G* nor its negation are provable, so *G* is undecidable, and this is not to say that it is the truth of *G* that leads to a contradiction. More precisely, what leads us to contradiction is the belief that, as *G* is true, *G* has also to be provable, for *G* is provable only if its negation is provable too. Thus, if *G* is provable, arithmetic would be inconsistent. And the fact that *G* is true leads us, not to its provability, but precisely to its unprovability.

As a whole, it is hard to believe that Russell might have developed this line of misinterpretation, which openly starts from a confusion between truth and provability. It is true that Russell referred to Gödel's incompleteness theorems as "a new set of paradoxes" (1945), and also as "a new set of puzzles" (1950), but is also true that he added: Gödel "proved that in any formal system it is possible to construct sentences of which the truth or falsehood cannot be decided within the system", which seems to show that Russell did not confuse truth and provability. Therefore, even in case that Russell actually did not understand Gödel's results, I think that the letter to Henkin is not enough to show that he thought that those results lead to the inconsistency of arithmetic. Rather, in writing the lines I have quoted above, he must have thought simply that his (and Whitehead's) lack of care in *Principia* about consistency was due to his belief that if arithmetic can be reduced to a few logical axioms, then any contradiction we find in further development must be due to the falsehood of at least one of those axioms, which can hardly be the case if the axioms are evidently and absolutely true. And this can be further interpreted as showing that after all Gödel's results have little to do with ordinary mathematics, which as is evidently true, is then consistent.

On the other hand, Russell recognized in the same letter to Henkin that although Gödel had shown that their carelessness about consistency was a mistake, he thought, "that it must be impossible to prove that any given set of axioms does *not* lead to contradiction". And Henkin, in his response, wrote precisely that Gödel gave a rigorous demonstration that this intuitive belief was absolutely true. The fact that Russell's original letter, no matter how oldfashioned it can be, contained interesting elements, was also admitted by Gödel himself, when a copy of the letter reached him (thanks to A. Robinson according to their correspondence now in Gödel's *Nachlass*). Gödel wrote to Robinson on July 2, 1973, the following: "Russell evidently misinterprets my result; however he does so in a very interesting manner". Unfortunately, he did not clarify the particular misinterpretation nor the interesting element he had in mind, so we cannot know whether he shared Henkin's and Dawson's interpretation.

In addition, Gödel did not take into consideration the fact that in the 1971 edition of his Schilpp volume (pp. xviii f), Russell wrote a very little-known addendum which I think can be regarded as his final word on this matter. The main parts of that addendum are: (i) a correct exposition of the essentials of the incompleteness result; (ii) the insistence that this result cannot be regarded as a fatal objection against the truth of mathematical logic, and especially against the absolute and general truth of every-day arithmetic: "It is maintained by those who hold this view that no systematic logical theory can be true of everything. Oddly enough, they never apply this opinion to elementary every day arithmetic"; (iii) a further statement of his belief that proving the lack of contradictions among the consequences of a system is impossible, "since the number of consequences of any given set of axioms is infinite".

I think this text makes clear that my interpretation seems to be correct, as it shows that bis 1963 reference to every-day arithmetic was intended merely to reinforce our intuitive confidence in the truth of self-evident elementary axioms. Also, the further appearance of his old belief about the impossibility of proving consistency shows once again that for him at least one of Gödel's results was not that surprising. At the same time, it shows the importance of an "empirical" viewpoint for Russell's philosophy of mathematics, which can doubtless be regarded as an important point of coincidence with Gödel's.

I come now to (ii), namely the assertion that according to Russell Gödel's "new puzzles" can be solved through the theory of types. The main evidence in print can be found in Russell 1959, where we read that he was convinced of his suggestion in the introduction to the *Tractatus* that a hierarchy of languages may allow us to dispose "of Wittgenstein's mysticism [his thesis of the ineffability of logical form], and, I think, also of the new puzzles presented by Gödel". This is not clearly referred to Russell's theory of types, in spite of the fact that the hierarchy of languages was doubtless inspired by it. However, I have found further evidence in an unpublished letter of April 6, 1960, to "Monsieur Morgenstern" (now in the Russell Archives), where we read: "As regards your contradiction derived from the principle that a law cannot be applied to itself, I think that the doctrine of types will deal with it as well as with Gödel's problems." Then our task has to be first to make more precise the sense in which for Russell there was something in Gödel's arguments, probably related to self-reference, which could perhaps violate the theory of types, and the sense in which this has to do with the hierarchy of languages.

According to the only author I know who has taken into consideration in print Russell's 1959 passage, Diaz 1975, Russell's mistake must have been the belief that every circular expression (an expression which is referred to itself) is a nonsense. And, as Gödel's proposition *G*, although circular, does not violate the theory of types, through the well-known device of representing the syntactical metatheory in the theory itself (the Gödel numbering), then there is not any nonsense in Gödel's argument. But I think this line is very open to doubt. For one thing it is very hard to accept that Russell merely believed that there cannot be acceptable self-reference (i. e., "this statement has five words"). Second, it is even harder to believe that Russell did not understand the fact that *G*, the famous undecidable statement, was a metamathematical statement asserting that *G* is not provable and, at the same time, an arithmetical statement constructed through the arithmetization of syntax in the same language. Besides, there is a passage in Russell 1940 where he shows his mastering of some of the possibilities of Gödel-numbering to refer to some statement without repeating it. Therefore, there must have been something else in Russell's mind.

Nicholas Griffin wrote to me about that particular point: "I think he would have thought that Gödel-numbering violated the sort of prohibitions on self reference that type theory was intended to impose. At the same time, he probably would not have felt confident that his version(s) of type theory or those actually articulated by anyone else were actually capable of ruling out Gödel numbering.“ But as I said above, Russell himself was well aware of the correct use of Gödel-numbering when he wrote that his theory of types (or his distinction between several languages) was able to deal with "Gödel's problems". I think in so doing he was probably thinking not of his classical theory of types in the sense of 1908 or 1910, but rather in an ideal "extended theory of types" as it was somehow intended in the Introduction to the *Tractatus*. If this were so, then he would be thinking of the possibility of languages of different orders to solve Wittgenstein's mysticism and Gödel's problems, in the precise sense that he was interpreting Gödel as falling down in the "paradox" of expressing syntax (Wittgenstein's "form") in the same language (apart from Gödel-numbering in itself). So, the ultimate source of his misinterpretation of Gödel might lie in his impression that Gödel's methods supposed a violation of the requirement that the syntax of a language, although it can be expressed (against Wittgenstein), this can be done only in a higher order language (against Gödel). If this is correct, then we need now to examine Russell's difficulties in handling the difference between theory and metatheory.

4.2 II

Russell was unable to handle the distinction between theory and metatheory.

On this point several accusations can be found in the literature: that Russell was unable to distinguish between logic and metalogic, between system and metasystem, between axioms and rules of inference, etc. (see the articles by Kreisel, Goldfarb, and van Heijenoort). As for textual evidence, §17 from 1903 is usually mentioned, as well as the corresponding place in 1910. There Russell says that for primitive (indemonstrable) propositions the method of proving independence is not applicable, because this method consists in supposing an axiom to be false and deducing the consequences. But "all our axioms are principles of deduction; and if they are true, the consequences which appear to follow from the employment of an opposite principle will not really follow, so that arguments from the supposition of the falsity of an axiom are here subject to special fallacies". Also, in 1910, p. 92, Russell wrote that we have the same difficulty with primitive ideas.

The ultimate reason for that attitude seems to be Russell's belief in the absolute truth of primitive propositions, and in their also being rules of inference. As for the main consequence, it is doubtless Russell's empirical approach to logic, in which the reasons for accepting axioms are inductive, and the method of discovering them "regressive", namely that which is based on their true consequences. However, although Russell's non-distinction between axioms and rules of inferences is undeniable, there exist places in which he tried to distinguish between true logical laws and rules of inference (i. e. general logical schemes) in a way that seems to me to depend on some application of metalinguistic distinctions, and also on deeper philosophical reasons. For instance, in the same §17 of 1903 Russell states the axiom "A true hypothesis in an implication may be dropped, and the consequent asserted", then he adds: "This is a principle incapable of formal symbolic statement, and illustrating the essential limitations of formalism". A little later this sentence is somewhat clarified, and Russell admits that with that he was trying to solve Carroll's puzzle in "What the tortoise said to Achilles", which may lead us to an endless regress of more and more premises.

Then Russell introduces (§38) a distinction between mere conditional implication, lacking genuine assertion and truth, and the connective "therefore", which, in asserting the truth of the antecedent, would be able to avoid the endless regress: only in that way could the hypothesis actually be dropped and the conclusion asserted by itself. I think with that Russell was somehow referring to something similar to what today we call a scheme or rule of inference, in the sense that it would sum up many cases of true logical laws. If this is so, Russell would be solving the puzzle with the standard distinction between language and metalanguage. And this is very interesting, because one of Russell's main reasons in being careful with these kinds of problems was his caution with Bradley's paradox against relations (and with Carroll's puzzle) that requiring the acceptance of ever more explicit relations doubtless leads us to a similar relational endless regress.

I always wondered whether someone sometime dared to ask Russell frankly about his difficulties in applying metalinguistic distinctions. Fortunately, when I asked Irving Copi about this, he told me he actually asked Russell in 1939 about it. Copi explained the results this way:

> Here I have in mind the difference between Primitive Propositions '*1.1 Anything implied by a true elementary proposition is true. Pp' and '*1.3 $q. \supset . p \vee q$.' At my first conference with Russell I inquired if the difference wasn't that one was a Rule *for* the language (and therefore *in* the metalanguage) whereas the other was a statement *of* the language and therefore *in* the object language. Russell answered 'Yes' and explained that in 1910 he and Whitehead hadn't yet 'got on to that way of talking'. He was not at all that comfortable at having to reply thus to my questions.

The fact that Russell referred to the whole matter as a new way of talking seems to me to be a sign that for him the new talk did not solve the deep question of the relationship between levels of language, as far as it can be related to the danger of some form of endless regress. But at any rate it can be also a sign that for him the distinction was well known, although obviously the later jargon was not.

Finally, I would like to insist that it was precisely Russell who introduced a hierarchy of languages of higher and higher order in his introduction to the *Tractatus,* and did so precisely in order to overcome Wittgenstein 's thesis that we cannot refer with sense to logical forms. According to Russell we can actually refer to a logical form present in a language, but only from a language of higher order. Ultimately, however, this actually amounts to very little, for although Russell accepted and sometimes used metalinguistic distinctions, he still seemed to accept Wittgenstein's dictum that we cannot express the syntax of a language in the same language, so he had great trouble in accepting Gödel's devices.

Gödel's famous 1931 results were possible precisely because of his new idea according to which, as truth and provability are so different, while the truth of the propositions of a language cannot be expressed in the same language (recall the paradox of the Liar), provability can, without thereby falling into any paradox. Thus, while for the early Wittgenstein the syntax of a language cannot be expressed at all, and for the later Russell it can only be expressed in a higher language, Gödel discovered how it can be done even in the same language, with which he showed the way to overcome one of the deepest beliefs of most of the members of the Vienna Circle, who had been influenced by Wittgenstein on this particular point. Perhaps Russell's strange position is reflected in this rather obscure passage from "On verification" (1937):

> It is true, as Tarski (*Der Wahrheitsbegriff in den formalisierten Sprachen,* Lvov, 1935) and Carnap *(Logical Syntax of Language,* Kegan Paul, 1937. –German, 1934) have proved, that in any given language there are things that cannot be said, but they can be said in a language of higher order. To say something about what cannot be said at all is not-necessarily self-contradictory, but there is no reason known to me for supposing that there is any actual significant statement of this sort.

If so, Russell, although accepting the possibility of using higher order languages without contradiction, continued to feel uncomfortable about its usefulness. Perhaps his difficulty in accepting Gödel's devices was a consequence of his suspicious attitude toward the actual possibility of saying something really "significant" about what cannot be said, and Gödel's celebrated undecidable proposition seemed to be a perfect example of this situation.

4.3 III

Russell's logicism collapsed after Gödel's incompleteness results.

The canonical position about the impact of Gödel's results on Russell's logicism is well represented by Henkin's "Are mathematics and logic identical?". First, *Principia mathematica,* as any other system incorporating arithmetic, is necessarily incomplete; second, if that system is consistent, no sound proof of this fact can be produced in the system. The conclusion is that any effort to establish a close relation between logic and mathematica is a failure. Thus, the only sense in which we can continue to speak of reducibility of mathematics to logic is by understanding logic as embracing set theory, as was Russell's original intention. But "no fixed system of axioms for set theory is adequate to comprehend all of those principles which would be regarded as 'mathematically correct'". Hence: "A true understanding of mathematics must involve an explanation of which set-theory notions have 'mathematical content', and this question is manifestly not reducible to a problem of logic, however broadly conceived" (Henkin 1962).

However, some arguments can be made in defense of Russell's original logicist intentions, even after Gödel's results (I follow here Sternfeld 1976 and Rodríguez-Consuegra 1991). First, although Russell spoke of reducibility of all mathematical propositions to logic, he did so in an informal sense; namely, without explicitly stating that the logistic thesis entails the provability or disprovability of *every* well-formed mathematical statement, while Gödel did not actually show any absolutely undecidable statement, but only a relative one. Second, Russell did not maintain a conception of logic according to which logical axioms are limited in content and can be clearly distinguished from mathematical axioms. Thus, there is no justification for saying that logicism is necessarily limited to a certain set of axioms which are not rich enough to found mathematics, as the writings by Carnap and Quine have later shown. Therefore, "since the logistic thesis presumed to show only that mathematics was derivable from logic, the fact that from no single axiom set can all the mathematical truths be proved reveals the limitations of mathematics which is perfectly consistent with the logistic thesis as similar)y limited to what is possible mathematically" (Sternfeld 1976). And that means that what was actually a catastrophe for Hilbert, who explicitly looked for completeness and decidability, is not necessarily so for Russell.

Besides, the fact that logicist reduction cannot be complete does not affect its truth, namely the fact that the instances of reduction which were actually done remain quite correct and are philosophically interesting in themselves (as was already pointed out by Putnam 1967). I think that is why Russell, who did not assume any completeness goal, maintained a much more empirical attitude to mathematics, and insisted that he was always convinced about the existence of undecidable axioms, like infinity and choice, which have to be accepted mainly because of their usefulness and "empirical" acceptable consequences, while he always thought that the consistency of an interesting axiom system cannot be proved at all (see above).

I have no room here to refer to the literature about saving certain "weaker" forms of logicism (Carnap, Hellman, Musgrave, Sainsbury), but I think as a whole they all are enough to show that some form of philosophical logicism, which was the only interesting one for Russell, can still be maintained, by simply accepting and incorporating Gödel's results as mere mathematical limitative results. I also think that it would be sufficient for Russell himself, who was never able to understand that, after Gödel, many logicians (although not certainly Gödel himself) regarded his magnum opus, *Principia mathematica,* as only of historical interest. Let me then finish by quoting the following letter, in which Russell's attitude can contribute to partially explain his strange position before Gödel (1968, p. 224):

> In particular, I am grateful for the nice things you say about *Principia Mathematica* and about me. The followers of Gödel had almost persuaded me that the twenty man-years spent on the *Principia* had been wasted and that the book had better been forgotten. It is a comfort to find that you do not take this view.

Acknowledgements I am very grateful to Ken Blackwell (the Russell Archivist, McMaster University) for the help and information provided, as well as to Elliot Shore (Institute for Advanced Study, Princeton), for his permission to study—and quote from—the Gödel correspondence. My thanks are also due to Irving Copi, John Dawson, Solomon Feferman, Ivor Grattan-Guinness, Nick Griffin and Greg Moore, who helped and encouraged me about the topics of this paper. Finally, I thank the organizers of this wonderful conference on the Philosophy of Mathematics for their kind invitation, and also an anonymous referee for useful suggestions.

References

Black, M. 1933 *The nature of mathematics.* London: Routledge.

Dawson Jr., J. W. 1984 "The reception of Gödel's incompleteness theorems". In Shanker 1988. (Originally in PSA, vol. 2.)

Carnap, R. 1934 *The logical syntax of language.* London: Routledge, 1937.

Diaz, E. 1975 *El teorema de Gödel.* Pamplona: EUNSA.

Gödel, K. 1931 "On formally undecidable propositions of *Principia mathematica* and related systems I". In Gödel 1986 (originally in German).

Gödel, K. 1986. *Collected Works. Volume I. Publications 1929–1936,* edited by S. Feferman *et al.*, New York and Oxford: Oxford University Press.

Goldfarb, W.D. 1979 "Logic in the twenties: the nature of the quantifier". *Jrn. Symb. Log.* 44: 351–368.

Hellman, G. 1981 "How to Gödel a Frege-Russell: Gödel's incompleteness theorems and logicism". *Noûs* 15: 451–468.

Henkin, L. 1962 "Are mathematics and logic identical?" *Science* 138: 788–794.

Kreisel, G. 1980 "Bertrand Arthur William Russell, Earl Russell. 1872–1970". *Biographical Memoirs of Fellows of the Royal Society,* vol. 28: 583–620.

Ladriere, J. 1957 *Les limitations internes des formalismes.* Louvain: Nauwelaerts.

Musgrave, A. 1977 "Logicism revisited". *British Jrn. Phil. Sci.* 28:99–127.

Putnam, H. 1967 "The thesis that mathematics is logic". In Schoenman 1967: 273–303.

Rodríguez-Consuegra, F.A. 1991 "El logicismo russelliano: su significado filosófico". *Crítica 61*: 15–39.

Russell, B. 1903 *The principles of mathematics.* Cambridge: Cambridge University Press.
Russell 1908 "Mathematical logic as based on the theory of types". *Amer. Jrn. Maths.* 30: 222–62.
Russell 1910 (with A.N. Whitehead) *Principia mathematica,* vol. l. Cambridge: Cambridge University Press.
Russell 1937 "On verification". *Proc. Arist. Soc.* 38: 1–20.
Russell 1940 *An* inquiry *into meaning and truth.* London: Allen & Unwin.
Russell 1945 "Logical positivism". *Polemic* 1: 6–13.
Russell 1950 "Logical positivism". *Rev. des mathématiques.* Rep. in 1956: 367–382.
Russell 1956 *Logic and knowledge.* R.C. Marsh (ed.). London: Allen & Unwin.
Sainsbury, R.M. 1979 *Russell.* London: Routledge.
Schilpp, P. 1944 (ed.) *The philosophy o/ Bertrand Russell.* La Salle, Ill.: Open Court, 1971.
Schoenman, R. 1967 (ed.) *Bertrand Russell. Philosopher of the century.* London: Allen & Unwin.
Schrin, M. 1976 (ed.) *Studien zu Frege/Studies on Frege,* 3 vols. Stuttgart u. Bad Canstatt: Fromman.
Sternfeld, R. 1976 "The logicist thesis". In Schrin 1976, vol. 1: 139–160.
van Heijenoort, J. 1985 "Système et métasystème chez Russell". Logic Colloquium '85. North-Holland, 1987.
Wittgenstein, L. 1922 *Tractatus logico-philosophicus.* London: Kegan Paul.

Departamento de Lógica, Historia y Filosofía de la Ciencia
Universidad de Barcelona

Chapter 5
Gödel's First Works, 1929–1936: Mathematics Without Philosophy?

Abstract This paper is a critical Essay Review of Kurt Gödel, *Collected Works. Volume 1. Publications 1929–1936.* Editor-in-chief Solomon Feferman. New York and Oxford: Oxford University Press, 1986, xvi + 474. First, I will describe the content of the book from the point of view of Gödel's works, making some comments on the way they have been organized for publication (II). Then I will describe the different sections other than the introductory notes, adding more comments about their usefulness (III). Finally, I will concentrate on the introductory notes, first to indicate their general features (IV), and second to describe and discuss, often very critically, the content of some of them, precisely those devoted to the philosophically most significant of Gödel's writings included in this volume (V-IX).

Keywords K. Gödel · Logic · Mathematics · Philosophy

Mathematics Subject Classification Primary 00B60 · Secondary 00A30

5.1 I

This is a very special book. It contains the official and definitive German and English versions of one of the most famous and influential papers in the entire history of logic, written by one of the great geniuses in the intellectual history of the human race. Besides, it contains all the remaining publications by Gödel in the period covered, in both languages (when necessary). It is the first volume of a long series which intends to critically edit all of Gödel's publications, together with the most significant unpublished writings. In addition, the team of six editors is impressive: the three mathematical logicians Solomon Feferman, Stephen Kleene and Robert M. Solovay, and the three historians of logic and mathematics John W. Dawson, Jr., Gregory H. Moore and the late Jean van Heijenoort, all of them working under the auspices of the Association for Symbolic Logic. They have been in charge of the completion of Gödel's publications, the general introduction, the chronology, most of the introductory notes, the textual notes, and the selection of the references (not

F. Rodríguez-Consuegra, *Russell, Gödel, Tarski*, Studies in Universal Logic,
https://doi.org/10.1007/978-3-032-08945-8_5

to mention other minor sections which will be described below). Moreover, there are other important names in charge of particular introductory notes, including B. Dreben, A.S. Tœelstra, W.D. Goldfarb, W.V. Quine, J. Webb, R.L. Vaught and R. Parikh. We have then before us a very important book, and to write a critical study of it is a complex task.

At this point, I must state what will be my main criticism of this edition, which I have tried to capture in the title I have chosen: it seems to me that the philosophical motivations and implications of Gödel's results are not elaborated on enough. This conviction depends upon a general thesis which I think, if true, is very important: that Gödel was mainly a philosopher searching for mathematical results to illustrate, if not to prove, the truth of a philosophical position, apart from the mathematical interest of those results. That philosophical position was a twofold kind of Platonism, i.e. ontologically, the belief in the existence of separate and transcendent abstract entities, and epistemologically, the belief in the existence of a human intuition allowing us somehow a direct access to those entities. Also, I would suggest that in spite of this particular criticism, this book is going to be the official, canonical and of course indispensable source for the study of Gödel's works for many years. This is appropriate, because the work is very competently done and contains the essential instruments which the Gödel scholar, or any other person interested in logic, mathematics or, more partially, philosophy, could possibly need.

5.2 II

The writings by Gödel which are published here can be divided into three categories: (i) major writings; (ii) minor writings; (iii) reviews. The major writings are the following. First, the article from 1930 on completeness of first order logic, together with the original dissertation from 1929, which is extremely interesting, not only because of the several modifications which appear only in the article, but also because the dissertation contained a fascinating philosophical and—mildly—polemical introduction which was later deleted from the actual publication. This is the first time this introduction is published, so the reader interested in historico-philosophical matters will doubtlessly study it carefully. Second, we have the star of the volume: the bilingual version of "On Formally Undecidable Propositions of *Principia Mathematica* and Related Systems I" (1931). This has been printed many times, but it is nice to have it at last together with the rest of its natural neighbors, like the several summaries and notes which Gödel published on several occasions, and, remarkably, Gödel's contribution to the Königsberg discussion of 1931 containing the first announcement of the incompleteness results. Third, there are the Princeton lectures from 1934 (first published in 1965) containing some developments of the incompleteness results, in particular a section entirely devoted to some relationships between those results, the paradoxes and the theme of truth, which was unfortunately missing from the original paper of 1931. Lastly, we have the paper from 1933 in which Gödel proved that—in a special sense—intuitionistic arithmetic is not really narrower than classical arithmetic (see above). As we shall see below, every one of these writings can be interpreted as having important philosophical motivations and implications which are not always adequately treated in the corresponding introductory notes.

The minor writings are very different in nature, extension and importance. Among them, we can find a few which are closely related to the ones I have mentioned above, that is, closely related to the themes of completeness, incompleteness and intuitionistic logic and arithmetic. Also, there are two from 1932 and 1933 devoted to the decision problem, as well as one from 1936 on the theme of the length of proofs. Incidentally, it has to be said for the decision problem that, although Gödel proved that certain classes of formulas of first order logic are decidable, his extension of the same argument to the predicate calculus with identity was shown to be false in 1984 by Goldfarb, who, fortunately, is in charge of the corresponding introductory notes. This is the only technical mistake Gödel made in print, as far as we know. If we add that some years later he withdrew at least one paper from publication when other mistakes were found (see p. 27), we can arrive at the comforting conclusion for the ordinary logician and philosopher that he was, after all, human too. Also, there is a series of three short notes from 1932 and 1933 on several themes of the propositional calculus, one of them on the possibility of precisely defining the expression "*p* analytically implies *q*", which seems to me to show Gödel's—presumably philosophical—interest in analyticity. Finally, we have five short notes on geometrical themes from 1933, which, although hardly interesting for the usual Gödel scholar, can be seen as a further sign of his impressive intellectual capacities.

The third and last category is constituted by the reviews. We have here a long series (more than thirty) of short reviews of books and articles, some of them very interesting in principle because of the obvious connection of the authors reviewed to Gödel's results, e.g. papers by Heyting, von Neumann, Skolem, Quine, Carnap, Church, Hahn and Hilbert. The reviews are however rather disappointing for the ordinary reader, who may hope to find some of Gödel's *opinions* about some of the fundamental ideas of those authors. For instance, as soon as I saw the book for the first time I immediately read the reviews devoted to Carnap and Hilbert, with the hope of reading something really fascinating, and even devastating, on Carnap's logical syntax of language (as he tried to present a conventionalist alternative to any Platonistic account of mathematics) and Hilbert's formalist program (as he tried to provide complete formalizations of mathematics, and perhaps dispense with some of the traditional abstract entities needed by the Platonist). Unfortunately, Gödel completely avoided any opinions and limited himself to a short description, as objective as possible, of the content of the several papers. This however is interesting, for together with the fact that he deleted the introduction to his dissertation from the article he wrote for publication, it provides a clear sign of one of the more remarkable features of Gödel's behavior: his extreme caution in expressing philosophical views. This was probably due to his highly idiosyncratic personality, which made him avoid philosophical polemic, particularly in an era when nominalism and empiricism became what he once described as "the prejudices of the time."

As a whole, most of these materials can be regarded as practically new, as many of the notes and reviews were unknown and have been rediscovered only very recently by John Dawson, who not only organized the Gödel *Nachlass* in 1982–84 and wrote interesting papers on Gödel and the reception of his results, but also is now writing what I am sure is going to be an impressive biography of Gödel [see Chap. 8 of *this* book]. One completely new item is the first printing of Gödel's original dissertation. Now we have for the first time a splendid, and probably complete, anthology of the period covered.

Let me call attention to a couple of critical points. All bibliographical references are made by the convenient, and highly historically informative style of author/year of publication. This is fine for books and papers by authors other than Gödel himself. But in the case of Gödel's own writings, a glance at the contents of the book may mislead the reader into seeing the several references as being somehow equivalent to each other, while actually there really are major works intermingled with short notes. It would have been more informative to adopt another system at least for major works, leasing the year/letter style for minor papers, notes and reviews. As for the overall decision to adopt the year of publication/presentation, although correct, some exceptions might perhaps have been allowed. For instance in the case of *1931* (the paper containing the incompleteness theorems which appeared in January 1931) and *1931a* (the contribution to the Königsberg symposium, which took place in September 1930), the reader is led to believe that the second is later than the first, and this is true if we mean that the second was *published* in 1931, after the first had already appeared. But it is also true that in *1931a* Gödel implicitly referred to, and actually announced his results of *1931*; so, in this case, choosing the date of the *presentation* would have avoided certain problems.

5.3 III

I come now to sections other than the Gödelian materials themselves and the introductory notes. A useful and friendly preface is provided by Professor Feferman in which he tells us the history of the project. There we learn, among many other things, that the encouragement of Gregory H. Moore and Jean van Heijenoort was of "pivotal importance" for its development. A useful section "Information for the Reader" follows, providing some comments explaining the criteria which have been chosen for organizing the introductory notes, references, textual notes, translations, etc. (I shall say something about these criteria in this and the next section). Then, after the copyright permissions, a note of gratefulness to the late Julia Robinson, and the "Contents," we arrive at the first major section of the book, Professor Feferman's thirty-six-page essay, "Gödel's Life and Work."

In short, I think this essay is the best short account of the global Gödel available in print, so it can be used as a splendid introduction to every aspect of our theme. The first part, devoted to Gödel's life and career, is preceded by the legendary words of Adele Gödel (Gödel's wife, who never studied philosophy, let alone mathematical logic) after attending Gödel's highly technical Gibbs Lecture: "Kurtele, if I compare your lecture with the others, there is no comparison." The short biography is very accurate and, at the same time, very interesting for the reader. Also, Feferman's obviously sympathetic attitude does not led him to hagiography. He does not avoid discussing the more obscure aspects of Gödel's personality, for instance his more or less frequent nervous depressions in the thirties, his hypochondriacal tendencies and his fascination with demonology.

One particularly significant feature of Gödel's life is also clearly noted: the fact that he devoted himself almost entirely to philosophy from 1943 onwards (when he was only 37 years old!), that is, after his main results in metamathematics and set theory were achieved, and after he failed to prove the independence of the axiom of choice and the generalized continuum hypothesis from the standard axioms of set theory (which was proved by Cohen in 1963). Usually, we would describe as a philosopher a person who devotes 13 years to a particular activity and the rest of his life, 25 more years, to philosophy. The fact that Gödel is not so regarded is because he published highly influential results only in mathematical logic and published very little on his philosophical "results." This is one of my arguments supporting the thesis that Gödel was mainly a philosopher, who devoted a few years to a certain branch of a particular exact science, mainly because he expected to find exact results to be presented as "proofs" that his philosophical doctrines were true.

The second part of Feferman's essay, devoted to Gödel's works, thought and influence, is divided into four sections. In the first one, all the materials of the first two volumes are grouped, very judiciously, into thirteen systematic categories, each one of them provided with excellent summaries of their contents. The second section groups the materials which are extant in the *Nachlass* into four categories: unpublished manuscripts, individual lecture texts, lecture notes, and notebooks. Unfortunately, there are no summaries here, as the reader is directed to "a succeeding volume," while no detailed prospective of that, or other succeeding volumes, is offered. Section 5.3 gives a fine and clear exposition of Gödel's philosophy of mathematics. However, Feferman's treatment of Gödel's doctrines consists of seeing them as "emerging" from the publications (the metaphor is used twice: pp. 22 and 30), which will probably be interpreted by most readers as if Gödel's philosophy of mathematics were some kind of "secondary" product implicit there. Feferman is clearer in Sect. 5.4, where he cites Gödel's realist position as a strong motivation and an important part of the reason for his success (p. 32). I think that Gödel's philosophy of mathematics was only part of a consistent whole, and that it can be interpreted not only as a fortunate motivation, but also as the main objective of his intellectual life. In short, and very roughly: the belief in transcendent entities is not only good because it leads us in the right direction to find metamathematical results, but these results are somehow the proof that those entities exist.

There are only two more sections remaining: the useful Gödel chronology by Dawson, immediately before the first of Gödel's papers, and the accurate "Textual Notes" (containing the necessary references to the copy-texts, the paginations and the replacements which the editors have introduced in the original texts), immediately after the last paper.

The bibliographical references and the index are doubtlessly very important in evaluating the usefulness of a work like this. The list of references (about whose author or authors we are told nothing) has been apparently constructed according to a threefold set of criteria (p. vi): Gödel's bibliography, and items referred to by Gödel or by the editors or other authors of the introductory notes. So, we would expect that every item in the list, if not by Gödel himself, was referred to by some member of the team somewhere in the book. But this is by no means the case. Thus, the index contains many items as appearing only in the list of references. Russell's works are a good example: while the corresponding entry (p. 447) offers us ten titles, we learn in the Index (p. 471) that only three of them are actually referred to in the body of the book. Since readers do not have at their disposal any criterion to determine why certain titles appear and why others do not, they can conclude that a certain degree of arbitrariness is present.

More concretely, there are a few significant omissions, which I have chosen more or less at random (there are *many* more). First, it was the Spanish edition of Gödel's (1981) *Obras Completas,* edited by Jesús Mosterín, which was the first attempt in any language to collect Gödel's publications. Second, is what I think can be described as the most famous first introduction to Gödel's incompleteness theorems: Nagel and Newman's (1958) *Gödel's Proof.* (I recently told Professor Dawson about these two omissions. He told me that the first one was not intended, while the second was the result of an explicit decision, as the famous booklet is simply a poor introduction.) Third, the very influential article (Lucas 1961), which gave rise to a whole series of replies, and is now indispensable in evaluating some of the philosophical implications of Gödel's incompleteness results for the problem of the nature of the human mind. Fourth, the massive (Hofstadter 1979). I think that these last two omissions are further consequences of the rather unfair treatment of philosophical matters which is present in this work (I will return to that below).

I have only one criticism to make of the index: it is an index only of names. It is perfectly legitimate and understandable, but the value of the book for the scholar would have been much greater had she/he been able to follow the track of certain key terms throughout the very dense xvi + 474 pages.

5.4 IV

The introductory notes to Gödel's papers and reviews try to provide (p. vi): (i) the historical contexts; (ii) explanations of the contents; (iii) discussions of further developments; (iv) critical analyses (admittedly, only in some cases). The Editors say that no uniformity has been imposed, although every note has been accepted by them as a whole. However, the notes meet these common purposes in varying degrees. Since the authors are highly competent in their field, the results are usually extremely good in spite of the lack of uniformity. However, as for the general degree of difficulty of the notes, I think that very few readers will be able to understand them without, at the same time, being able to understand the papers themselves. So, from the point of view of the "explanation of the contents," the notes seem to me to have been composed rather for the specialist than for the general reader interested in Gödel coming from other fields of science or philosophy. It is in this sense that the four purposes have been planned more around the interests of mathematical logicians than of philosophical logicians or simply philosophers. Thus, consideration of the historical *philosophical* context, the *philosophical* implications of the contents, further *philosophical* developments, and critical *philosophical* analyses, are unfortunately almost absent from the introductory notes, apart from some allusions, or the use of a few standard philosophical terms and labels.

From the point of view of the reader, it is very easy to know at first glance whether one is looking at an introductory note or one of Gödel's writings: in the first case there is always a vertical line on the outside which runs the full length of the page. Also, every note begins with the common form "Introduction to... (the key/s to the paper/s)," and we are told that every one of Gödel's writings has the corresponding introductory note. However, some readers will perhaps be puzzled in trying to find the note to a particular paper or review, as in some cases the editors have grouped, under a particular introductory note, the material corresponding to several writings. This is not a serious problem; it is only a matter of some inconvenience. But the inconvenience increases as: (i) the papers and reviews are printed chronologically, so the materials covered by the same introductory note may be located in very different places in the book; (ii) Gödel's intellectual production in these years consisted mainly of many short publications a year (e.g., from 1932 we have *1932, 1932a, ...1930o,* that is 16 items!), so sometimes it is difficult to find quickly what is needed.

The Editors must have considered the case where the material covered by the note is not located immediately after it. We may read, for instance, a short "[The introductory note to... can be found on page..., immediately preceding...]" before the first line of the paper in question. However, the treatment is not always uniform, as there are instances where common introductory notes simply say almost nothing about some of the writings. One example of that is [Gödel *1932b*], "On Completeness and Consistency" (p. 235 ff), which is only mentioned in the "Introductory Note to *1930b, 1931* and *1932b*" (p. 126 ff), by Professor Kleene, who devotes only two and a half lines to it (p. 126). The reader especially interested (this was my case), is forced to go to the index or the bibliographical reference to learn something else about that particular paper, and what can be found is only the place and time of publication. Perhaps this is all the information relevant to this paper, but I cannot help being convinced that a much better system would have been: "a paper, an introductory note."

5.5 V

From here on, I will concentrate on five of the introductory notes devoted to the five papers or notes which are philosophically more important, without forgetting that the importance of an idea in history of philosophy is measured by the influence exerted by its direct or indirect influence on the intellectual period considered.

I will begin with the introductory note to Gödel's dissertation of 1929 and related materials (p. 44 ff), by Burton Dreben and Jean van Heijenoort. Let me first say clearly how much I admire this note, which can be described as historically superb. Its first section is devoted to the historical background of the idea of completeness, that is: the Frege-Russell inability to imagine a non-universalist and comprehensive logic; the absence of the notion of formal system in Peirce-Schröder; the work by Post and Bernays demonstrating the semantic completeness of the propositional calculus, and the first appearance of the same question for quantification theory in Hilbert-Ackermann.

Missing here is only some philosophical explanation for the reason why Frege-Russell thought it impossible to look at logic from outside it. I think the crux of this question was their belief that any attempt to do so leads to unsolvable paradoxes belonging to the same family as Bradley's paradox against relations. Thus, to escape from logic presupposes defining its main notion, that of logical form, which is a kind of relation, in such a way that logical form appears to be a genuine concept (a "term") related to other concepts. But as soon as we say that we are forced to explain the relation between forms and non-forms, which leads us to other forms, that is, to an infinite regress. As we shall see below, a parallel philosophical problem appears in relationship to Gödel's incompleteness results and the question of the expressibility of truth and syntax in the same language.

The study of the introduction to the dissertation is philosophically the most interesting section. The authors accurately explain Gödel's main argument against the formalist belief that the consistency of an axiom system implies the existence of the corresponding mathematical concept: "it assumes that no formula in the notation of the system can be proved to be undecidable" (p. 49), that is, it assumes that every problem is solvable. Then, after criticizing Gödel's example of an axiom system for the reals with two non-isomorphic models for not mentioning the distinction between first and second order logic (which presumably means that the argument is not correct, although this seems to be only suggested; see [Feferman 1984, 551, footnote 7], who describes the argument as "inconclusive"), Dreben and van Heijenoort point out that Gödel's criticism of the formalist belief goes further than Brouwer's, for it is not only that we cannot simply assume the solvability of every mathematical problem, but also "that we may be able to prove the unsolvability of some mathematical problem" (p. 50), with which the incompleteness results seem almost to be announced.

However, the authors do not mention the fact that it was precisely Gödel's completeness theorem which, in demonstrating that if an axiom system is consistent then it has a model, proved that consistency is somehow equivalent to existence. So, his reluctance to admit this equivalence can be interpreted as meaning simply that we cannot assume it *before* having obtained the completeness result (on that point see [Feferman 1984, 551 f]). Thus, the following step towards proving incompleteness can be seen as pointing out the definite limits of consistency from the ontological point of view: before incompleteness was proved, the equation of consistency with existence could be still maintained. Perhaps this may help us to understand Gödel's final decision to delete the introduction from the printed version, a problem which the authors leave open (as Feferman did in his [1984, 552]).

The main section of this introductory note is the technical one devoted to the proof of completeness itself, whose precise progress is completed by comparisons with some of Skolem's former ideas (where important letters by Gödel are quoted), as well as with Herbrand. A significant summary of these comparisons is the following: "let us consider the sequence: Frege-type formal system, proof procedure, set-theoretic validity. Skolem connects (at least implicitly) the second and third terms of the sequence, Herbrand the first and second, Gödel the first and third" (p. 55). The section finishes by making some comments on further developments by Gödel himself and Henkin, and so does the last short section. As for the *introductory* nature of this note: I would have liked to read somewhere a few simple paragraphs emphasizing the intuitive gist of the proof and the main standard consequence for logical truth and provability, very roughly like the following, which can be found in standard introductions to logic.

As for the basic idea of the proof, we can say that the completeness of quantification theory (that every one of its valid formulas is a theorem) is proved by showing that every valid first-order formula can be reduced to a valid propositional formula from which it can be derived. Then, as the propositional calculus had already been proved to be complete (by Bernays), every valid propositional formula is also provable, so every first-order formula (which can be derived from one of the former) is provable too, and first order quantification theory is complete. In short, what Gödel proved was that first-order logic is complete if propositional logic is complete.

As for the standard interpretation of the completeness result, we can say that it shows that for first order languages a formula is logically true iff it is logically provable, i. e. truth is equivalent to provability, which means that the corresponding proof procedures are complete. This can be seen as philosophically relevant, especially if we express the same idea in other —and later— terminologies (and always for first-order logic): logical consequence is equivalent to derivability; semantic consistency is equivalent to formal consistency; every semantically valid formula (true in every system) is syntactically demonstrable (derivable with no premise); in short: semantics is equivalent to syntax.

5.6 VI

Gödel's contribution to the Königsberg conference of 1930, published in 1931 after his incompleteness results, was the next natural step after his former unpublished criticisms against formalism that we have discussed above. John Dawson's introductory note to that contribution comes from his excellent paper on the same topic (Dawson 1984), in which he provided a wider commentary, as well as the English translation of the contributions by other participants (Hahn, Carnap, von Neumann).

The most interesting point of Gödel's contribution, apart from containing the official announcement of his completeness result, is the continuation of his dissertation criticisms against consistency of an axiom system as a criterion for existence of a mathematical concept, which is now based upon an implicit announcement of the incompleteness results. However, the idea that is expressed here is rather against consistency as a criterion for the truth of every theorem provable in the system. Dawson summarizes the argument by saying that "contrary to Carnap, Gödel argues against adopting consistency as a criterion of adequacy for formal theories," for one could perceive, finitarily and contentually, "that a statement provable within some consistent formal system is nonetheless false," as "one can give examples of contentually true propositions that are unprovable within the formalized framework of classical mathematics" (p. 197). This summary, essentially correct, leaves unexplained several interesting points, doubtless because for Dawson they are too obvious. But I think for the general reader these points should have had more elaboration.

First, the allusion to Carnap. I think it is correct, providing Carnap's philosophy of mathematics at that time can somehow be identified with that of Hilbert and his followers, against whom Gödel's remark is explicitly directed (he speaks of "the formalist view," p. 201). For Hilbert, a proof of consistency for a system would guarantee the truth of every theorem provable within it. (In this he directly contradicted Frege, for whom the true situation was just the reverse: a set of true axioms is already trivially consistent, so there is no need of any proof of consistency.) If we can prove that there can be a consistent system from which we can derive a false theorem, then the proof of its consistency is not enough. And this is what Gödel says can be proved through his argument.

Secondly, it seems to me that the link between the fact that a false statement can be provable within a consistent system, i.e. that of classical mathematics (which we take to be so), and the fact that contentually true propositions can be unprovable within this system is not emphasized enough. I think this link is established by the conclusion to Gödel's contribution: if there is some contentually true proposition which is unprovable in classical mathematics, then "if one adjoins the negation of such a proposition to the axioms of classical mathematics, one obtains a consistent system in which a contentually false propositions is provable" (p. 203). Therefore, we can add, consistency does not guarantee truth, as consistent systems can contain false statements.

In this context it may be worth recalling Russell's early strong opposition to consistency as an ultimate criterion for truth, which appeared in several places in his work, for instance, against Bradley's coherence theory of truth, and against "implicit definitions" as being able to guarantee the existence of the concept supposedly defined by them. Contrarily, for Russell (i) consistency is not enough, as there can be consistent systems constituted by false statements; (ii) we need first prove the existence of the concept involved.

One of the philosophical morals of Gödel's global argument can be stated this way: if consistency does not guarantee truth or existence, we cannot create objective concepts simply by constructing axiom systems, i.e., concepts exist or not on their own. A second moral could be: any attempt to interpret mathematics in a formalistic—then somehow conventionalist—way is hopeless, precisely because we by no means can forget mathematical intuition, nor replace it by a proof of consistency which can be understood in an algorithmic, mechanical way. Needless to say, this is one of the main lines of the philosophical implications of the incompleteness results. So, there should be some reference to it in this introductory note, as it was the first place in which Gödel made public some of those results, and he did it in a full philosophical context.

5.7 VII

I come now to the kernel of Gödel's work in this period: his celebrated paper of 1931 containing the incompleteness results for which Professor Kleene wrote the exquisite introductory note. It would be difficult to find a person better qualified for this task.

There is a clear summary of Hilbert's program (a program based on the hope that consistency and completeness are attainable for formalized mathematics), then a short account of Gödel's incompleteness results showing the impossibility of meeting both requirements at the same time. This is followed by a brief exposition of the central ideas of Gödel's first theorem. Next the exposition of Gödel's paper begins, which is structured according to its four original sections, and constituted the bulk of the introductory note. Finally, three pages are devoted to further developments in the field, particularly several attempts to provide finitary consistency proofs.

As a whole I think the note is technically perfect, although it is very difficult to comprehend if we regard it as a first "introduction" to the theme. From an historical viewpoint I think the situation is unfortunately different. It is true that Professor Kleene makes useful references to Gödel's development, starting from the discovery of the undefinability of arithmetical truth in arithmetic (p. 127), as well as Gödel's first treatment of the question at the Königsberg meeting (pp. 135, 137). Also, he devotes a footnote (p. 137) to refer the reader to the literature on the reception of Gödel's results. However, there is no discussion of the possible problem of priority for what now is known as "Tarski's theorem" of the undefinability of truth in the same language, which is particularly needed as Kleene seems to accept that Gödel already had the idea in 1930. This is needed especially since Kleene himself devotes some interesting remarks to interpret Gödel's rather cryptic footnote 48a, by writing

(p. 135): "Implicit in this remark is that the adjunction of higher types to a formal system permits one to define the notion of truth for that system, then to show that all its provable sentences are true, and hence to decide the sentence shown in Theorem VI to be undecidable in the system" (see below for further discussion of this point).

Moreover, as for Gödel's development of the incompleteness idea, we do not find any treatment of the historico-philosophical information available in the literature at the time when the book appeared (mainly Feferman (1984) and Dawson (1984, 1985)). For instance, there is no allusion to Finsler, the only precedent to Gödel. It is true that he was mistaken, especially as he thought absolute formal undecidability could be proved, but I think a few lines about him would have been useful for the reader to have another instance of the historical fact that original ideas always have some sort of precedent. As for the reception of Gödel's results, I am convinced that a simple footnote to the relevant literature is not enough. Some treatment of the main problems forestalling full acceptance is also necessary.

From the philosophical viewpoint things are, in my opinion, not much better. I do not have enough space to devote even a few lines to every missing point I think would have been relevant here, so I shall only mention what I think are the most important of them:

- the actual historical reception of Gödel's results from the viewpoint of at least a few well-known names: Hilbert, Brouwer (see (Wang 1987), Russell (see Rodríguez-Consuegra (1992)), Wittgenstein (see Shanker 1988, 155–256));
- Gödel's conviction of the objectivity of mathematical truth, as opposed to mere provability;
- implications for logicism: impossibility of a unique global account of formalized mathematics;
- possibilities of partial reconstructions of the logicist program by assuming Gödel's results;
- implications for intuitionism: Gödel's interest in giving a proof acceptable to the intuitionists, as he had available earlier a general proof of incompleteness which was not acceptable (letter to Zermelo of 1931; see Grattan-Guinness (1979);
- implications for Platonism: are Gödel's results an argument to vindicate Platonism? See Myhill (1952);
- implications for realism: are Gödel's results an argument to vindicate philosophical realism? See Dummett (1963);
- implications for conventionalism: can Carnap's syntactical program of 1934 be refuted by Gödel's results, though it was explicitly constructed by starting from them? See Bohnert (1975);
- implications for the analytic-synthetic question: could Gödel's results have helped to give rise to Quine's famous attack against the analytic-synthetic distinction? In particular, can those results be interpreted as having proved that some synthetic knowledge is possible, or at least that mathematics cannot be analytic? See Copi (1949) and Körner (1967);

- are Gödel's results relevant for the problem of the nature of the human mind, especially concerning algorithmic possibilities? (see above, Sect. 5.4, and below, Sect. 5.9).

I am fully aware of the limitations imposed on an introductory note, and of the fact that often mathematical logicians are not very fond of philosophical questions. In particular, I know that Professor Kleene has written elsewhere (1976/78, 77) that he does not consider himself qualified to give an account of Gödel's work in philosophy. But I cannot help feeling that since this is the official edition of Gödel's works, the introductory note to such an incredibly rich and influential paper should have been explicitly treated some of its most important philosophical implications, regardless of the view that Gödel was considered to be only a mathematical logician.

5.8 VIII

Gödel's interest in intuitionism hardly needs to be emphasized. As I pointed out before, one of the main motivations for his particular way of proving formal undecidability was that the proof would be accepted by intuitionists. In the period covered by this book, Gödel wrote several short papers which have to do with intuitionism. I think two of them are relevant to understanding Gödel's philosophical motivations, because they provide further evidence that most of the time he may have regarded mathematical results also as the basis for his philosophical convictions. The papers were written in 1933 and are entitled "An Interpretation of the Intuitionistic Propositional Calculus" and "On Intuitionistic Arithmetic and Number Theory."

For the first paper, Professor Trœlstra writes in the corresponding introductory note that it contains an interpretation of intuitionistic propositional logic in an extended system of classical propositional logic. Then, after describing the details and further developments, he adds: "For Gödel, the interest of his result presumably lay in the fact that it gave for IPC an interpretation which was meaningful also from a non-intuitionistic point of view" (p. 299). Regarding the second paper, Gödel first showed that classical propositional calculus is a subsystem of the intuitionistic one, and then that something similar holds for arithmetic and number theory. On the implications of this thesis, Professor Trœlstra, who is also the author of the corresponding introductory note, only writes: "Gödel concludes the note with the observation that his results show that intuitionistic arithmetic is only apparently weaker than classical arithmetic" (p. 283).

But we can ask, why was Gödel so interested in this sort of reducibility of classical to intuitionistic formal systems? To provide some answer to this question is relevant for intuitionism as well as for formalism and logicism (which were more inclined to use classical logic and arithmetic) and are mainly *philosophies of mathematics*. A part of the answer is provided by Gödel's own remark at the end of the last paper indicating that "an intuitionistic consistency proof for classical arithmetic and number theory" is involved in this result (p. 295). This is true, we can add, because the reduction proves any theorem of classical arithmetic to be *classically* equivalent (that is why intuitionists can reject the result!) to a theorem of intuitionistic arithmetic, so we can no longer think that within intuitionistic arithmetic we can feel surer of avoiding contradictions than within classical arithmetic. This is perhaps too obvious, but once it has been explicitly noticed, one can immediately understand that the intuitionists' extreme caution in the fully philosophical search for a secure basis for mathematics, could be overcome, at least to some degree, through a technical result. If so, this can be interpreted as another instance of Gödel's hope of providing technical arguments for philosophical polemics.

5.9 IX

To complete this critical study, let me say something about Gödel's Princeton lectures of 1934. Professor Kleene again is responsible for the excellent introductory note to those lectures, which he personally attended and, together with J.B. Rosser, prepared the notes they took for publication.

Kleene explains, one by one, the technical content of the nine sections of the text, as well as the "Postcriptum" from 1964, showing that the lectures cover more territory than the original paper of 1931, especially regarding some improvements in the proof that every primitive recursive function is representable in the system, and the definition of a new and important mathematical idea, that of a general recursive function. Kleene goes on to describe further developments, where the names of Church, Turing, Rosser, Tarski, Mostowsky, Robinson and Kleene himself have introduced to establish the relevant links with Church's thesis (that every effectively calculable function is general recursive); Church's undecidability theorem; Turing machines and computability; Rosser's replacement of Gödel's ω-consistency; Tarski, Mostowsky and Robinson's work on undecidable theories, and Kleene generalizations of Gödel's first theorem.

Three points, all related to the philosophical issues which I have referred to above need to be made. First, I find Kleene's treatment of section 7 of Gödel's text entirely insufficient. He only says that it contains a discussion of the relation of his arguments to the paradoxes, where Gödel credited Carnap and Tarski for the general self-referential lemma and the undefinability of truth in the formal system. As for crediting Tarski, I think some discussion of the possible priority is necessary in view of the fact that it may be said that Tarski's theorem of the undefinability of truth in the same language can be credited instead to Gödel. Some of the grounds for believing this to be the case are the following (for the full historical evidence see mainly Grattan-Guinness (1979), Wang (1987) and Coffa (1987): (i) it was known to Gödel as early as 1930; (ii) it was proved by Gödel in 1931, at least before Carnap (and perhaps before others); (iii) it was proved in the letter to Zermelo of 1931; (iv) Tarski arrived at the discovery of the theorem only after having read about Gödel's undecidability results; (v) even so, Tarski did not include *clearly* the undefinability of arithmetic truth in arithmetic in the 1933 monograph —where Theorem I of 85 is usually, and mistakenly, cited on this point— but rather in the Postcript from 1935; (vi) Tarski himself, in the footnotes to his famous monograph, recognized his dependence upon Gödel, and so did Gödel, in an unpublished letter to Burks of February 6, 1964.

Secondly, section 7 contains something extremely valuable from the historico-philosophical viewpoint: Gödel's first proof in print (Princeton 1934) that truth cannot be expressed in the same language, which is immediately applied to arithmetic, is Gödel's first statement in print of the heuristic argument for the existence of undecidable formulas. The argument, which already appeared in his letter to Zermelo of 1931, reads as follows: "So we see that the class α of numbers of true formulas cannot be expressed by a propositional function of our system, whereas the class β of provable formulas can. Hence $\alpha \neq \beta$ and if we assume $\beta \subseteq \alpha$ (i.e., every provable formula is true) we have $\beta \subset \alpha$, i.e., there is a proposition A which is true but not provable. $\sim A$ then is not true and therefore not provable either, i.e., A is undecidable" (p. 363). I think this great argument deserved a more complete treatment.

Gödel's use of his technique of arithmetization in reconstructing the same argument for undecidable propositions, to show that a particular undecidable proposition can actually be constructed through the expressibility of the syntax of a language in the same language (1931), seems to me of the utmost philosophical importance too. Thus, he solved a very old philosophical problem from the time of Bradley and Frege, for whom only the universal logical language was inescapable. Russell tried to overcome the problem through his theory of types, but Wittgenstein rejected it through his dictum that we cannot speak about language at all, as there is really only one language; therefore, Russell wrote in his introduction to the *Tractatus* that there is at least one escape: one can speak of a language by constructing a metalanguage. Then Gödel (and Tarski) showed that we can express the syntax of a language in the same language without contradiction. Thus, the only thing we cannot express in a language is its semantics, which requires a metalanguage.

Finally, Gödel's "Postcriptum" of 1964 contains the first statement in print of his full philosophical conviction: that the incompleteness results, together with the implications of the ideas involved in the concept of Turing machines, "do not establish any bounds for the powers of human reason, but rather for the potentialities of pure formalism in mathematics" (p. 370). Gödel had developed this statement in his Gibbs lecture of 1951, as well as in his contributions to Wang (1974). This is a further point which, I am convinced, requires a more adequate treatment. To summarize my underlying argument, which has been applied throughout this study: if some philosophical ideas were true and important for Gödel, they have to be made known to everyone interested in any facet of his thought.

Acknowledgements I am grateful to John Dawson for all the general information he has given me on the global project of Gödel's *Collected Works,* as well as on some of the points I mention in section III. Also, my thanks are due to Ignasi Jané, who read an earlier version of this critical study and made remarks which led me to introduce several improvements, both in content and in exposition. Finally, I am indebted to Francine Abeles, the Book Review Editor, for improving my English.

References

BOHNERT, H. 1975. *Carnap's logicism,* J. Hintikka (ed.), *Rudolf Carnap, logical empiricist,* Dordrecht, Reidel.

COFFA, A. 1987. *Carnap, Tarski and the search for truth,* Noûs, 21, 547–572.

COPI, I. 1949. *Modern logic and the synthetic a priori,* Journal of Philosophy 46, 243–245.

DAWSON Jr., J. W. 1984. *Discussion on the foundation of mathematics,* History and Philosophy of Logic 5, 111–129.

DAWSON Jr., J. W. 1985. *The reception of Gödel's incompleteness theorems,* in P.D. Asquith and P. Kitcher (editors), *PSA 84: Proceedings of the biennial meeting of the Philosophy of Science* Association (East Lansing, Michigan, Philosophy of Science Association, 1985) vol. 2, 253–271. Reprinted in [Shanker *1988,* 74–95] and in T. Drucker (editor), *Perspectives on the history of mathematical logic,* (Boston/Basel/Berlin, Birkhäuser, 1991), 84–100.

DUMMETT, M. 1963. *The philosophical significance of Gödel's theorem,* Ratio 5, 140–155.

FEFERMAN, S. 1984. *Kurt Gödel: conviction and caution,* Philosophia Naturalis 21, 546–562.

GÖDEL, K. 1981. *Obras Completas,* edited by J. Mosterin (Madrid, Alianza); second, really "complete," edition, 1987.

GRATTAN-GUINNESS, I. 1979. *In memoriam Kurt Gödel. his 1931 Correspondence with Zermelo on his incompletability theorem,* Historia Mathematica 6, 294–304.

HOFSTADTER, D. R. 1979 *Gödel, Escher, Bach: an eternal golden braid,* New York, Basic Books.

KLEENE, S. C. 1976/78. *The work of Kurt Gödel,* Journal of Symbolic Logic 41, 761–778; *Addendum,* 43, 613. Reprinted in [Shanker *1988,* 96–114].

KÖRNER, S. 1967. *On the relevance of post-Gödelian mathematics to philosophy,* I. Lakatos (editor), *Problems in the philosophy of mathematics* (Amsterdam, North-Holland), 118–133.

LUCAS, J. R. 1961. *Minds, machines, and Gödel,* Philosophy 36, 112–127.

MYHILL, J. 1952. *Some philosophical implications of mathematical logic,* Review of Metaphysics 6, 165–198.

NAGEL, E. & NEWMAN, J. R. 1958. *Gödel's proof.* New York, New York University Press.

RODR1GUEZ-CONSUEGRA, F. A. 1992. *Russell, Gödel and logicism.* Paper read at the Wittgenstein Symposium on Philosophy of Mathematics, Kirchberg, Austria, August 16–22, 1992. [Reprinted in *this* vol. ch. 4]

SHANKER, S.G. 1988. (ed.), *Gödel's theorem in focus,* Beckenham/N.Y., Croom Helm.

WANG, Hao 1974. *From mathematics to philosophy,* New York, Humanities Press.

WANG, Hao 1987. *Reflections on Kurt Gödel,* Cambridge, Mass., MIT Press.

Departamento de Lógica, Historia y Filosofía de la Ciencia
Facultad de Filosofía, Universidad de Barcelona

Chapter 6
Gödel's Last Works, 1938–1974: The Emerging Philosophy

Abstract This paper is a critical Essay Review of Kurt Gödel, *Collected Works.* Volume II. *Publications 1938–1974.* Editor-in-chief Solomon Feferman. New York and Oxford, Oxford University Press, 1990, xvii + 407. My viewpoint will be that of a philosopher, so I will make comments mostly on the philosophical implications of—or the philosophical theses explicitly maintained in—this set of Gödel's works, as well as on the way in which those implications are taken into consideration in the corresponding introductory notes.

Keywords K. Gödel · Logic · Mathematics · Philosophy

Mathematics Subject Classification Primary 00B60 · Secondary 00A30

6.1 I

This is the second volume of this impressive series of Gödel's works. Its general characteristics are the same as those of the first volume, which appeared in 1986 with the same editorial team. Since I already reviewed those characteristics in my earlier essay review (this journal, 3, (1992), 58–74 [*this* vol. Chap. 3.]), I will concentrate here on the particular content we are offered now. The reasons for adopting a philosophical viewpoint, already stated in my former review, can be summed up here: while Gödel's technical works have been well studied and have exerted massive influence in the logico-mathematical development of the second two thirds of this century, his philosophical ideas have been rarely taken into consideration by philosophers. Also, I think that much of Gödel's technical results were obtained mostly in search of logico-mathematical support for his philosophical beliefs. In this connection this second volume of his published works is really fundamental, as it was only in this period that Gödel decided to make public some traits of his philosophical position, which had been maintained at least from the mid-twenties, and to which he devoted,

F. Rodríguez-Consuegra, *Russell, Gödel, Tarski*, Studies in Universal Logic,
https://doi.org/10.1007/978-3-032-08945-8_6

as a whole, much more time than to his technical works, in spite of the fact that he persistently refused to publish his purely philosophical writings.[1]

For convenience, the content of this volume can be divided into six parts:

1. The preliminary notes and abstracts plus the definitive work dealing with Gödel's proof of the relative consistency of the axiom of choice (*AC*) and of the generalized continuum hypothesis (GCH) with the standard axioms of set theory (1938–1940);
2. The article on Russell's mathematical logic (1944);
3. The two versions of the article on Cantor's continuum problem (1947–1964);
4. The technical and philosophical writings on relativity theory (1949–1952);
5. The two versions of the paper on finitary mathematics, the second printed here for the first time (1958–1972);
6. Several remarks and notes.

In the following, I will comment on each of these items, particularly on the respective introductory notes, which is what is really new in this volume, and their treatment of the philosophical questions involved.

6.2 II

Gödel's proof of the relative consistency of the axiom of choice and the generalized continuum hypothesis with the usual axioms of set theory can be seen as the last of his great results, although in this case Gödel was unable to finish the task of proving not only relative consistency, but also independence, which was done by Cohen in 1963. The introductory note to these writings, by R. Solovay, seems to me historically and technically superb. After explaining the historical antecedents of Zermelo's axiom of choice and Cantor's continuum hypothesis, the different axiomatizations of set theory are briefly reviewed in order to point out the model-theoretic work which had been done in set theory before Gödel, that is, to the ideas used by Fraenkel, Skolem, von Neumann and Ackermann on consistency and independence of various axioms.

Then Solovay goes on to a description of Gödel's proof of relative consistency, using constructible sets whose hierarchy was seen by Gödel as a "natural prolongation" to transfinite levels of the ramified theory of types of *Principia Mathematica*. The intuitive idea, i.e. Gödel's "inner model" method, is clearly stated: "he described

[1] Gödel's most elaborated philosophical unpublished manuscripts, written in the fifties, are going to appear in volume III of this collection, mainly the Gibbs Lecture (Providence, R.I., 1951) and "Is mathematics syntax of language?", versions III and V (which were written in 1953–1959 as Gödel's intended, but never actually submitted, a contribution to Schilpp's Carnap volume, finally appeared in 1963). A personal reconstruction and a long historico-philosophical introduction by this reviewer of the Gibbs Lecture, as well as of versions II and VI of the other manuscript mentioned, with a foreword by W.V. Quine, is forthcoming in Spanish (Kurt Gödel, *Ensayos inéditos,* Barcelona: Grijalbo-Mondadori, in print [actually appeared in 1994]) and is in preparation in English. [Kurt Gödel, *Unpublished philosophical essays*. Basel, Boston and Berlin, Birkhäuser, 1995].

a certain collection of sets, called the *constructible* sets, and was able to prove (in axiomatic set theory without the axiom of choice) that each of the axioms of set theory holds in the domain of constructible sets. He also showed that *AC* and *GCH* hold in this domain. From this it follows easily that if the axioms of set theory became inconsistent after adjoining *AC* and *GCH,* then they must already have been inconsistent without these new axioms" (pp. 6–7). Thus, the gist of the proof consists first of defining a constructible universe, through a certain hierarchy of sets using arbitrary ordinals as the levels, then of showing that *AC* and *GCH* hold in this universe, showing that they follow from the "axiom of constructibility" (V = L; i.e. every set is constructible) and that this axiom holds in the constructible universe. Solovay's long last section is devoted to a description of the further work which has been done, both on constructible sets and on the various questions raised by Gödel's work.

However, from the philosophical viewpoint nothing is said. It is certainly true that this work by Gödel is perhaps the one whose philosophical implications are most difficult to point out, but I think that at least two questions should be noted as being philosophically relevant. First, to what extent can this *result* by Gödel be seen as providing arguments either for realism or for nominalism (or positivism, as Gödel himself used to say)? Second, can Gödel's *methods* be presented as some sort of illustration of his well-known realistic position?

As for the second question, Solovay has correctly cited (p. 8) the relevant reference to Gödel's letter of 1968 to Hao Wang, in which Gödel said that, as far as the continuum hypothesis (*CH*) is concerned, his proof of relative consistency was impossible to discover for constructivists because the corresponding ramified hierarchy was used there in a nonconstructivistic way. Solovay points out that the nonconstructivistic element "lies in the use of arbitrary ordinals as the levels in Gödel's extension of the ramified theory" (p. 8). But there is no attempt to develop the way in which this element can be related to Gödel's general realistic attitude. This is particularly unfortunate, because in the letter to Wang, Gödel tried to connect his methods in this work with his general philosophical position. In particular, I think that it is useful here to try to determine the different senses in which Gödel's position can be connected with his methods. In general, I think it can be said, perhaps at the same time, that Gödel's realistic thesis can be seen as: (i) a philosophical consequence of his technical results, (ii) a heuristic principle leading to them, and (iii) a philosophical hypothesis to be "verified" through them. Then, it might be said that the methods used in the relative consistency proof can be seen as proceeding from an attitude made possible by (ii). But this becomes difficult, when we try to do the same thing for (i).

Thus, I come to the first question, i.e. is Gödel's result here to be located on the side of realism? Gödel himself seemed to have great doubts regarding a positive response to this question. In Hao Wang's *Conversations with Gödel.*[2] Gödel said that this particular result was rather on the positivistic side, although he does not explain the precise sense in which these words should be interpreted. On the other hand,

[2] I have been able to read a copy of this unpublished work, which Professor Hao Wang has generously provided me. [See Chap. 9 of *this* volume.]

many would consider it more or less neutral as far as philosophical controversy is concerned. At any rate, it has to be recalled that for Gödel this particular result was to be seen as only the first part of the complete one: the independence *of AC* and *GCH* from standard set theory, as reached by Cohen in 1963. For this complete result we do have Gödel's detailed philosophical reaction, which was added as a supplement to the second version of his Cantor paper (1964). Let us then leave the matter for the corresponding section below.

6.3 III

Gödel's essay on Russell from 1944 is, we can say, the kernel of his emerging strongly realistic philosophy which he thought to be true at least from the mid-twenties. But it seems he did not dare devote room to it in print until he had finished his most impressive works in mathematical logic and set theory, probably because—among other things—he thought that his results might be used to support that philosophy. The essay is printed here with seven more pages of textual notes which were found in a series of offprints of the original paper, now in Gödel's *Nachlass.* Some of them are interesting inasmuch as they seem to show some of Gödel's doubts on a number of points. The introductory note has been very ably written by C. Parsons, who explains the origin of the paper, and gives a summary of its main contents in several sections, which is very useful as the original essay did not have any separate sections. The more important of these sections are devoted to the paradoxes and the vicious circle principle; Gödel's realistic position; the ramified theory of types, and Gödel's views on analyticity. However, along with the summary, Parsons adds comments on some of Gödel's views which is also useful, allowing the reader to state some links to certain philosophical problems. As a whole, this introductory note seems to me a good one, although I cannot help missing more historical information about the circumstances that could explain why Russell did not publish any reply to Gödel's paper in his Schilpp volume, as well as additional clarification of some of Gödel's extremely interesting philosophical views. I shall say something on these two points in the following.

Parsons says that Gödel sent in a first manuscript to Schilpp in May, and the final one in September, after some revision. He adds that Russell meanwhile had completed his replies to the other papers, so he decided not to reply to Gödel's but only to add a brief note at the end of the replies (which is transcribed here too) in which Russell says that the paper arrived when he "had no leisure to work on it". The problem is that one might think that Russell actually did see the May version (as suggested by Hao Wang in his *Reflections on Gödel*), and then decided not to reply, perhaps—the reader may conclude—for reasons other than the one he states in the note.

The study of the correspondence between Russell and Schilpp (now in the Russell Archives, McMaster University) decides the question completely: Russell was unable to see the May version because Gödel explicitly prohibited Schilpp to show it to him.

Besides, it shows that Russell was, during a period of some four months, eagerly waiting for some version of Gödel's paper to write the reply, even after he had finished with the rest of the replies. Finally, Russell was by then frantically busy: among other things preparing his departure for England. A further sign of his wanting to publish some reply is that in the 1971 edition of his Schilpp volume he published a little known, but very interesting short "addendum" to his replies which was almost completely devoted to Gödel. True, the addendum is no reply to Gödel's criticisms of 1944, but I think it can be regarded as some sort of compliment to Gödel, and, to my knowledge, it is the only important place in print in which Russell somehow gives us his view of Gödel's celebrated results.

The places where I miss more clarification of Gödel's philosophical ideas are the ones related to his well-known, although difficult to understand, realistic view, and to his thesis that mathematical propositions are analytic. As for Gödel's realism, Parsons makes useful comments on Gödel's main argument for realism: that the assumption of the objective existence of sets and concepts is as legitimate as the ordinary assumption of physical bodies to obtain a satisfactory theory of our sense perceptions (p. 106). It seems to me that when Gödel refers here to "physical bodies" he is also thinking of the "theoretical concepts" which physicists admit as long as they are convenient assumptions to obtain more satisfactory physical theories (the "inferences" in Russell's sense, as opposed to the "logical constructions"). But if so, to say that we admit certain objects to obtain a "satisfactory" system, either mathematical or physical, seems to me dangerously close to some sort of holistic position, for the role of an entity in a theory may in part depend on the relation of this theory to other auxiliary hypotheses or theories in a whole system of them. Thus, as holism is one of the main enemies of platonistic realism, a certain tension would appear at this point.

On the other hand, the expression "a satisfactory theory" seems to have some pragmatic connotations, especially when we realize that for Gödel the acceptance of certain axioms may depend on their "fruitful" consequences more than on their intrinsic truth, which is the only one to which our supposed faculty of mathematical intuition may lead us. If these unexpected connections are justified, then some sort of review of Gödel's concept of science, perhaps too dependent on the classical scheme of hypothesis-verification, would be necessary.

Regarding Gödel's view of analyticity, Parsons correctly analyzes the two senses of the word for Gödel (p. 116): mathematics cannot be tautologically analytic because it is undecidable, but it is analytic in the sense that its propositions are true in virtue of the meaning of their concepts. Besides, Parsons makes the important remark that this notion of analyticity is very close to the one which was maintained for the Vienna Circle, although he does not point out that Carnap's objection to Wittgenstein's notion of tautological analyticity in *Logical Syntax* is also similar to the one that Gödel made in his 1944 essay. However, no attempt is made at further analyzing Gödel's notion of meaning, for his underlying conviction that only by showing that mathematics is analytic can a realistic position be saved from empiricism. Gödel's unpublished manuscripts, in particular his Gibbs Lecture (1951) and his several versions of "Is mathematics syntax of language?" (1953–1959), can be used to throw some light on these problems, in particular on Gödel's analogy between mathematics and physics and on his thesis of analyticity. Parsons mentions the second of these manuscripts in a footnote (p. 117), but only mentions a "very brief examination", which is very unfortunate, for one of the more fruitful and legitimate purposes of the study of the unpublished material by an author is precisely to help in our understanding of her/his published writings.[3]

6.4 IV

The two versions of Gödel's essay on Cantor's continuum problem, published in 1947 and 1964, are reproduced in this volume, with an introductory note written by G. H. Moore, the historian of mathematical logic and set theory. The note is a model of clarity and accuracy from the historical and technical points of view. After explaining the historical background of the continuum problem from Cantor to Gödel's work before 1947, as well as the origin of Gödel's paper of 1947, Moore goes on to a description of its main content. In this section there is only one page devoted to—mostly a couple of quotations of—Gödel's Platonistic viewpoint, according to which Cantor's conjecture must be either true or false in itself, for its undecidability from the usual axioms of set theory is only a sign of the lack of further axioms which would describe more fully an objective, underlying reality, which has to be supposed in the same way as physical objects are assumed for physical theories. Then Moore adds three more sections, dealing respectively with the 1964 version, the later technical research affecting the problem (the more spectacular part of which is, of course, Cohen's results showing the independence of the continuum hypothesis from the standard axioms of set theory), and a very interesting survey of Gödel's several versions and drafts of an unpublished new paper on the problem, together with several highly relevant quotations from the correspondence of this time.

However, the philosophical treatment of the 1964 version is very limited, in spite of the fact that Moore himself says in the introduction to his note that this paper by Gödel contains no new technical results but "gives considerable insight into his philosophical views on set theory" (p. 154). That version contained a supplement mostly devoted to further developments of Gödel's philosophical objectivism for the existence of mathematical objects, but Moore give us in the corresponding place of his introductory note only a quotation and seven lines of comments. I have already pointed out the extreme importance of Gödel's philosophical position for Gödel himself, as well as some of the interesting problems underlying his analogy between mathematics and physics, so I think that this note would have been a good place to

[3] Parsons mentions some very interesting references where Gödel's philosophical ideas have been discussed, but there is nothing about the literature in which Gödel's philosophical fines are explored. In particular, nothing is said about P. Maddy's writings on mathematical realism, many of them already published before 1990. For a later treatment of that development Maddy's recent book is indispensable: *Realism in mathematics* (Oxford, Clarendon Press, 1990).

delve deeper into some of these problems, especially to try to establish some links to Gödel's technical arguments.

Yet even forgetting these problems, the five last paragraphs of Gödel's supplement (pp. 267–269) are philosophically so rich, so laden with epistemological and ontological implications, that there should have been at least a description of each of them, together with some comments connecting them to Gödel's other known arguments in print, as well as to some of Gödel's arguments in the unpublished manuscripts written between the two versions of the Cantor paper (specifically the Gibbs Lecture and the several versions of the unsubmitted essay on Carnap). In the end, if some resort to unpublished technical manuscripts is regarded as necessary to understand the implications of the published technical arguments, why not apply the same reasoning to the unpublished philosophical writings in connection with the published philosophical arguments? Unfortunately, the problem remains the same as I have repeatedly noted: Gödel's philosophical position cannot (should not) be put aside and be taken into consideration only in a secondary way in relation to his mathematical results and arguments: for him both were inextricably tied, and this in such a way that one can even maintain—as I do—that for him his philosophical beliefs were, by far, the more important, while his mathematical work was, somehow, a search for arguments which might be used in support of those deep beliefs.

6.5 V

I come now to Gödel's writings on relativistic cosmology and its philosophical implications. The reader will find here the two technical papers on Gödel's new solutions of the equations of Einstein's general theory of relativity of 1949 and 1952, in which Gödel introduced "rotating universes", as well as the philosophical one he published in the Schilpp Einstein volume of 1949. This paper, which also contains some footnotes that were added by Gödel to the German edition of 1955, was devoted to a discussion of the main consequences of his technical arguments: that the reality of time and change is rather illusory, as can be illustrated by the possibility of time travel which is allowed by the rotating universes. The introductory note to the technical papers is due to the physicist S.W. Hawking. In one page Hawking gives the reader some background to Gödel's solutions, some discussion of their relationship to Mach's principle, and extremely brief descriptions of their content, but no philosophical discussion is even attempted. However, a final note, added by the editors, mentions some detailed discussion of Gödel's cosmological ideas which can be found in the literature.

H. Stein wrote the introductory note to Gödel's contribution to the Einstein volume which contains some useful clarification of the kind of idealism that Gödel seems to hold in the paper, i.e. that change is not objectively real, although that does not necessarily imply the stronger thesis that time is mere appearance. Also, Stein discusses the possible relationships of Gödel's position to Kant's transcendental idealism, as well as the apparent discrepancy between the paper and an unpublished manuscript which can be found in Gödel's *Nachlass* dealing with similar questions: "Some observations about the relationship between theory of relativity and Kantian philosophy", from which some ideas are summarized.

I have only two criticisms to make. For one thing it seems to me a mistake not to have published here the manuscript just mentioned, together with a comparative study of it with the first one. The argument that these first two volumes are exclusively devoted to Gödel's *publications* might be acceptable, but—fortunately—that policy has not been consistently maintained: the first volume included Gödel's unpublished doctoral dissertation, and in this second volume the reader finds a second version of the *Dialectica* paper on finitary mathematics which had not been previously published. This was obviously done to shed light on the corresponding published material, so it seems to me that the same should have been done in this case. The second critical point concerns Stein's comments. In them there is nothing regarding the possible relationship between the technical results and their "philosophical implications", so the reader probably will come to believe that Gödel suddenly, and neutrally, "discovered" his new solutions, *then* realized that some important philosophical consequences could be drawn from them. However, another possibility might be more credible: it seems to me that we have here another case of Gödel's usual procedure of trying to demonstrate the truth of certain philosophical theses previously maintained: to explicitly *look* for scientific, hence undeniable, technical support.[4]

6.6 VI

I congratulate the editorial team for including in this volume the improved 1972 version of the *Dialectica* paper of 1958 on finitary mathematics, in spite of the fact that Gödel never decided to publish this second version. A. Trœlstra has been in charge of the introductory note, which runs as follows: a history of both papers (2 pages); a statement of Gödel's main aim (2 pages); a technical description and explanation of Gödel's methods and results (14 pages), and a study of further technical

[4] Two recent books are very relevant to the themes of Gödel's papers on relativistic cosmology and its philosophical implications: P. Horwich, *Asymmetries in time* (Cambridge, Mass.: The MIT Press, 1987), in which, among many other interesting things, Gödel's ideas are explicitly discussed, and the actual possibility of time travel is defended; P. Yourgrau, *The disappearance of time: Kurt Gödel and the idealistic tradition in philosophy* (New York: Cambridge Univ. Press: 1991), which contains a useful study of Gödel's unpublished manuscript mentioned in the text, as well as an attempt to relate Gödel's strange views concerning time and change to the rest of the philosophical theses he seems to have maintained. This second book appeared after 1991, so it could not appear in the bibliography, but the first one appeared in 1987, so it clearly should have been included, and even explicitly mentioned, especially because it contains a rich bibliography listing other papers which are relevant to a study of Gödel's cosmological ideas. I think this is another consequence of the unfair treatment of Gödel's philosophical ideas: while it is clearly understood that "further work" along these lines has to be explicitly taken into account, and even discussed, this is only done for his technical work.

related research (5 pages). As we can see from the number of pages devoted to each item, Gödel's aims in writing this paper have been considered only to a very limited extent, and only from a rather technical point of view, despite the explicit recognition by Trœlstra that, although Gödel himself said that in the 1972 version he was more concerned about the philosophical implications of these results, had already presented the 1958 paper "as a foundational contribution", that is, as one which is interesting mostly because of a philosophical gain. But philosophical gains are supposed to be justified from the viewpoint of some general philosophical program.

Trœlstra clearly describes Gödel's main aim as the attempt to justify classical systems as intuitively as possible, but without accepting Hilbert's restrictions to finitary mathematics, that is, to purely combinatorial mathematics admitting only finite, discrete and visualizable objects. Thus, although Gödel's incompleteness results force certain abstract notions to be admitted in consistency proofs, Gödel thought that Heyting's abstract notion of constructive proof had to be replaced by something more finitistic; hence his new system T, based on a certain notion of functional, which can technically replace Heyting's arithmetic, while preserving a more concrete, intuitive character. However, Trœlstra says nothing about the philosophical gain, apart from the reference to intuitiveness in general. It seems to me that this was important for Gödel, because from his realistic viewpoint it was essential to be able to show that we do possess an intuition which has some direct access to abstract notions, so the attempt to show this for the particular case of certain abstract notions needed in consistency proofs can be regarded as a partial development of his philosophical program.

From a more detailed point of view, Gödel's papers, especially the second one, contain many rich remarks, sometimes devoted to further elaboration of his notion of intuition which was so important in connection with his realistic program. Among them, the first two paragraphs of both papers, as well as notes b and c (p. 272) seem to me essential to understand Gödel's notion of intuition, in particular the way in which it has to be distinguished from the Hilbertian and the Kantian notions, which, in the end, depend upon a very restricted concept of finitary, discrete mathematics, which was unacceptable for Gödel's view of intuition, one being able to lead us to abstract notions, or, as he also says here, to meanings, structures or thought contents. But no attempt is made by Trœlstra to take into consideration such passages, which are not even explicitly pointed out to the reader. Thus, while those mainly interested in Gödel's technical ideas and methods doubtless will be satisfied with Trœlstra's note, those interested in Gödel's philosophical ideas, although glad to have the 1972 version now available, will continue to miss a philosophical study of its philosophical content, together with a comparison to the 1958 version.[5]

---

[5] Not even other philosophically relevant literature is mentioned. Here I am thinking of Hao Wang's *Reflections on Gödel* (Cambridge, Mass.: The MIT Press, 1987), which contains a brief but useful study of the philosophical implications of Gödel's paper, as well as of other of his papers with philosophical interest. It is really surprising that Wang's book has not been taken into consideration or even mentioned by any of the authors of the introductory notes. But what is really astounding to me is the fact that the book has not even been included in the bibliography.

6.7 VII

There are other remarks and notes which are also important from the philosophical viewpoint. Let us at least point them out. Gödel's remarks for the Princeton bicentennial of 1946 on absolute demonstrability and definability can be usefully regarded as "instances of the application to concrete problems of Gödel's realistic point of view", as Parsons has written in his introductory note to them (p. 149). Also, Gödel's "Some remarks on the undecidability results" (1972) contains rich comments on the philosophical side. The second of those remarks, although devoted to a description of another version of the first undecidability theorem, contains the only known definition of "analytic" – apart from the one included in the essay on Russell of 1944—which is very usefully discussed at length by S. Feferman and R. Solovay in the corresponding introductory note (pp. 289 ff). As for the third remark, although it comprises only 22 printed lines, it is so rich that J.C. Webb needed ' almost 12 pages to analyze its content, which deals mainly with Turing's notion of mechanical procedures in connection with the nature of the human mind.

6.8 VIII

As a whole, the book is absolutely indispensable for anyone interested in Gödel's ideas, or generally on the history and philosophy of logic and mathematics. As I have indicated, philosophers more interested in the philosophical implications of Gödel's ideas may be somewhat disappointed by the treatment that these ideas have received in some of the introductory notes, but the book is, at any rate, indispensable for them.

As for the third volume of this fascinating series which is being finished at the time I write this (March 1993), it will contain Gödel's most important unpublished manuscripts, most of them given as lectures or intended for publication, although never actually submitted. Regarding the editorial team, some changes have taken place. Unfortunately, Jean van Heijenoort died in 1986, although the second volume was mostly completed by then. The editors, however, have written: "His spirit will continue to animate all our work, and we have taken his standards as our own" (p. vi). Due to his work on Russell's *Collected papers,* Gregory H. Moore declined to continue after this second volume of the series, which is regrettable, given his ability. On the philosophical side, Charles Parsons and Warren Goldfarb have been added to the team and are in charge of the edition of some of Gödel's philosophically most important manuscripts. This is good news, so that readers should no doubt look forward to reading the forthcoming volume.[6]

Departamento de Lógica, Historia y Filosofía de la Ciencia
Facultad de Filosofía, Universidad de Barcelona

[6] My thanks are due to Francine Abeles and Roberto Torretti for improving the English of this paper.

Chapter 7
Gödel's Unpublished Manuscripts, 1930–1970: The Official Edition

Abstract This paper is a critical Essay Review of Kurt Gödel, *Collected Works. Volume III. Unpublished essays and lectures.* Editor-in-chief Solomon Feferman. New York and Oxford: Oxford University Press, 1995. xvii + 532 pp (For the preparation of the manuscript of this review resources provided by the grant DGICYT PS93-0220 are gratefully acknowledged. Also, I am grateful to Francine Abeles for her help with the English of a first version of this review.). As in my review of the vol. II, my viewpoint will be that of a philosopher, so I will make comments mostly on the philosophical implications of—or the philosophical theses explicitly maintained in—this set of Gödel's works, as well as on the way in which those implications are taken into consideration in the corresponding introductory notes by the editorial team.

Keywords K. Gödel · Logic · Mathematics · Philosophy

Mathematics Subject Classification Primary 00B60 · Secondary 00A30

7.1 I[1]

This book is the third of the important series that the Association for Symbolic Logic is devoting to critically editing Gödel's works and correspondence, published and unpublished. Yet there is an important novelty now; while the two former volumes contained all the published writings and reviews,[2] the present one includes most of the unpublished essays and lectures (in English, and also in the original German when relevant) which have been found in Gödel's *Nachlass*, most of which appear here

---

[1] For the preparation of the manuscript of this review resources provided by the grant DGICYT PS93-0220 are gratefully acknowledged. Also, I am grateful to Francine Abeles for her help with the English of a first version of this review.

[2] See my essay-reviews; this journal, 3 (1992), 58–74 and 4 (1994), 318–327. (This volume, Chaps. 3 and 5.)

F. Rodríguez-Consuegra, *Russell, Gödel, Tarski*, Studies in Universal Logic,
https://doi.org/10.1007/978-3-032-08945-8_7

for the first time.[3] Another important change is that some members of the editorial team have been replaced by other scholars. In particular, the late Jean van Heijenoort (deceased in 1986), the late Stephen C. Kleene (deceased in 1994), and Gregory H. Moore (who left to devote himself full-time to the Russell project) have been replaced by Warren Goldfarb and Charles Parsons.

As a result of being mainly a volume containing unpublished materials, other minor changes in the general pattern of former volumes have taken place as well. Thus, a preface by Solomon Feferman (the editor-in-chief), a brief overview of the Gödel *Nachlass* by John W. Dawson (pp. 1–5) and a report on Gödel's Gabelsberger shorthand by Cheryl A. Dawson (pp. 7–12) have been included. A further consequence has been the considerable expansion of the textual notes section (39 pp., against 13 pp. in vol. II). Unfortunately, the editors continue their policy of including only an index of names, but not one of concepts, which, in dealing with unpublished materials, would have been particularly useful here. The Gödel materials in this volume can be divided into two categories: technical and philosophical, although some of the technical writings include philosophical remarks. Yet while the technical are mostly expository and extend mainly from 1930 to 1941, the philosophical are clearly at the beginning of the volume and are a clear sign that from the early forties to the end of his life philosophical reflection was the most important intellectual activity for Gödel. Accordingly, I have organized this review by devoting the first section to the technical materials and the remaining sections to the four main philosophical writings appearing here. When I have something to say on the corresponding introductory notes, I do it in the appropriate section.

7.2 II

Most of Gödel's technical materials appearing in this edition are the texts of lectures given with the aim of explaining technical results or discussing possible ways of solution to certain problems. Thus, we can find, among others manuscripts dealing with more specialized points: lectures on the completeness of elementary logic; on the celebrated incompleteness theorems; on the consistency of the continuum hypothesis; on the sense in which intuitionistic logic is constructive, and—rather surprisingly—on some possibilities for continuing Hilbert's program in a revised form (whose original epistemological value is strongly defended).

The text of an invited lecture of 1933 in Cambridge, Massachusetts, in what was Gödel's first trip to the US, is particularly useful to provide us with a global view of "The present situation in the foundations of mathematics" (1933) by that time. In particular, Gödel, in a rather Russellian way, describes the foundational problem as

[3] Another book containing unpublished materials by Gödel appeared in the same year: K. Gödel, *Unpublished Philosophical Essays* (Basel/Boston/Berlin: Birkhäuser, 1995), edited by this writer. The Spanish edition appeared one year before: K. Gödel, *Ensayos inéditos* (Barcelona, Mondadori, 1994).

one divided into two parts: first to reduce the actual methods of proof to a minimum number of axioms and rules of inference; second to give a justification of these axioms. According to him, the first part can be easily solved by a simple theory of types without certain restrictions. The second is more difficult, for although we can deal with the formalism as if it were a symbolic game, it is when we attempt to attach some meaning to those symbols that serious problems arise: the notions of existence, of class, and the axiom of choice. The amazing point is that after discussing these difficulties, Gödel goes on to say that if we interpret the axioms in a meaningful way, then we are presupposing a kind of Platonism, "which cannot satisfy any critical mind and which does not even produce the conviction that they are consistent" (p. 50). Even after reading Feferman's analysis of this statement (pp. 39–40) the reader does not know what to do with it, although to say the least, it should be a sign of Gödel's constant difficult relationships with "public" and "private" Platonism. Another interesting lecture is the one of 1949 on rotating universes. The text is more or less the same as the published one (1949), and it contains new solutions to Einstein's field equations of gravitation, but as is usually the case with Gödel's manuscripts prior to publication, it contains more material than the one actually published.

An interesting example is Gödel's assertion that he arrived at the new solutions while he was working on the similarities between Kant's philosophy and relativity physics "insofar as in both theories the objective existence of a time in the Newtonian sense is denied. On this occasion one is led to observe that in the cosmological solutions known at present there does exist something like an absolute time. ... So one is led to investigate whether or not this is a necessary property of all possible cosmological solutions." (p. 274). If I understand this well, then it could be interpreted as a new argument in favor of a general thesis on the relations between Gödel's technical work and his global philosophical motivation of it.[4]

Another remarkable technical work by Gödel appearing here (although not for the first time) is his celebrated and much discussed logico-ontological proof of the existence of God, which is well-known at least since the early seventies, thanks to Dana Scott's first presentation in a seminar, after he was shown the proof by Gödel himself. The text is the strict presentation—in hardly one and half pages, pp. 403–404—of the proof in logical symbols, whose general theme is the transition from the essence of a being possessing all positive properties to her necessary existence. The introductory note, by R. M. Adams, is very useful, as it contains some historico-technical background, as well as a good discussion of Gödel's concepts and proof-steps, and a review of the relevant literature appeared between 1970 and 1991. Yet

[4] The present writer has been maintaining this general thesis in several publications on Gödel. See, for instance, my essay-review of vol. II of the Gödel series in this journal (4, 1994, 318–327 [Reprinted in *this* vol. Chap. 5]), especially concerning Gödel's work on rotating universes (where I already conjectured that Gödel's reasons for working on new solutions to Einstein's equations of gravitation must have been philosophical), as well as my introductory essay to the book cited in footnote 3.

I find that the not especially interested reader is going to miss not having a simple, intuitive presentation of the general structure of the proof.[5]

Finally, a still more famous text appears in this book: Gödel's never before published legendary paper entitled "Some considerations leading to the probable conclusion that the true power of the continuum is $\aleph_2$" (together with two other closely related writings, including an unsent letter to Tarski). This is the text which Gödel sent in 1970 to Tarski for submission to the *Proceedings of the National Academy of Sciences* that was never published. The paper is very short (not even three printed pages, pp. 420–422), but the introductory note by R. M. Solovay offers to the reader a full analysis of the axioms given by Gödel as well as of the alleged—but failed—proof, and some discussion of further possible work on similar lines and some open questions. Solovay was the referee who turned back the paper to Tarski saying that, "if the author were anyone but Gödel, I would certainly recommend that the manuscript be rejected" (p. 405), so his final remark providing some explanation for Gödel's errors is worth citing: "Ill health both made it seem urgent that his ideas be communicated to the world and made it impossible for him to carry out his usual scrupulously careful presentation and checking of the details" (420).

7.3 III

In the 1949 Schilpp volume on Einstein, Gödel published a short paper entitled "A remark about the relationship between relativity theory and idealistic philosophy". It was devoted to defending a particular kind of idealism (close to Parmenides and Kant) according to which relativity theory is not incompatible with the thesis that change is not objectively real. Among Gödel's papers, five manuscripts written between 1946 and 1949 have been found dealing with somewhat similar matters; two of them appear in this volume under the title "Some observations about the relationship between theory of relativity and Kantian philosophy." Yet they are not at all mere preparatory drafts of the Schilpp paper, which is more cautious and less extended in scope, but contain bolder and *more* extended arguments. According to them, although there is something important in Kant's philosophy which has been confirmed by relativity theory, some reinterpretation of that philosophy must be undertaken as a result of Einstein's theory.

I think the following passage by Gödel can be regarded as a good summary of his global position in these papers on the point that relativity theory is opposed to Kant (p. 244):

> A real contradiction between relativity theory and Kantian philosophy seems to me to exist only in one point, namely, as to Kant's opinion that natural science in the description it gives of the world must necessarily retain the forms of our sense perception and can do nothing else but set up relations between appearances within this frame. This view of Kant has doubtless

[5] See for instance, the one contained in Hao Wang, *Reflections on Kurt Gödel* (Cambridge, Mass., The MIT Press, 1988), p. 195.

> its source in his conviction of the unknowability (at least by theoretical reason) of the things in themselves, and at this point, it seems to me, Kant should be modified, if one wants to establish agreement between his doctrines and modem physics; i.e., it should be assumed that it is possible for scientific knowledge, at least partially and step by step, to go beyond the appearances and approach the world of things.

On the other hand, Gödel was convinced that relativity theory has confirmed Kant's doctrine that the natural world picture is essentially "subjectivistic" (p. 245). So, our view of the world and science should be a compromise between this confirmation and that rejection. The introductory note by H. Stein is clear and detailed, offering the reader both useful technical context and philosophical comments.

Two points seem to me worth recalling in connection with these two manuscripts by Gödel, one related to the content of Gödel's position, the other to his method of work. For one thing, it seems to be clear that Gödel is here maintaining a position very close to the one he was maintaining in his philosophy of mathematics. According to that position, while the objects of mathematics are objective, we do possess an intellectual intuition of them which, with some cultivation, is able to let us *see* their deep nature. Thus, Kant's philosophy should be amended in order to release itself of the limits of intuition based strictly on sense perception.

Secondly, the fact that Gödel wrote up to five extended and bolder manuscripts, when finally he published only a short, cautious and more general paper, seems to be due to his usual fear of making public his actual philosophical beliefs. Something similar took place with his paper for the Schilpp volume on Russell.[6] But this style tending to introversion, fear of controversy and the consequent increasing self-censorship was at its height in the series of papers on Carnap, as we shall see in the next sections.

7.4 IV

"Some basic theorems on the foundations of mathematics and their philosophical implications" was the title of Gödel's Gibbs Lecture, given at Brown University in 1951. It seems that Gödel had some doubts about publishing it because the essay, which is fully philosophical, contains a strong attack against Carnap's nominalism-conventionalism. So, Gödel finally decided not to publish it, probably because he was not completely satisfied with his own arguments, and was fearful of the controversy that would ensue.

The kernel of the lecture is the showing that mathematics is inexhaustible (through the exposition of the iterative concept of set and of Gödel's incompletability results), followed by the drawing of some philosophical consequences which, according to Gödel, can be adduced against Camap's conventionalistic viewpoint, then in favor of some sort of a platonistic philosophy of mathematics.

[6] See my review of vol. II cited in footnote 4.

The arguments against Camap's view that mathematics is a vast tautology are basically that the *explanation* that this is so cannot be given without resorting to the axioms of mathematics themselves, and that the corresponding *proof* would be also a proof of the consistency of mathematics, which is not allowed by Gödel's second incompleteness theorem. As for the arguments in favor of Platonism, they are mainly arguments against the view that mathematics is our own creation. For Gödel this cannot be so: first, because if mathematics were our creation we should see its inner more difficult intricacies with perfect clarity, which is by no means the case; second, because it is obvious that we cannot arbitrarily create the truth of the theorems, and finally, because (p. 314):

> If mathematical objects are our creations, then evidently integers and sets of integers will have to be two different creations, the first of which does not necessitate the second. However, in order to prove certain propositions about integers, the concept of set of integers is necessary. So here, in order to find out what properties *we* have given to certain objects of our imagination, we must first create certain other objects —a very strange situation indeed!

The text of the lecture is the accurate and painstaking reconstruction of the handwritten manuscript by Charles Parsons. This is a genuine editor's nightmare, for the original text is an almost inextricable tangle of revisions, deletions, crossed-out passages, footnotes (again revised, deleted, etc.) and additional notes to the text and to the footnotes (again revised, deleted, etc.), many of which cannot be properly located in the main text. The final result is admirable, although the editor does not give us all his criteria of reconstruction in the textual notes at the end of the volume. Yet a study of the final rendition of the text in comparison with the manuscript leads us to believe that he has kept everything which possibly could be inserted or linked to the main text. It is obviously a perfectly legitimate choice, although not the only one.[7]

The introduction, by George Boolos, is a model of clarity and usefulness for any serious reader. After a brief historical information and overview, he offers us a summary of Gödel's main arguments, together with some explanation of some of his concepts, and even some very convincing criticisms of Gödel's position. Yet those readers interested in locating Gödel's ideas of this essay in a wider historico-philosophical context will have to look elsewhere.

[7] Another choice, the one followed by this writer in his edition of this manuscript (see footnote 2), is trying to keep absolutely everything which can be actually read in the manuscript, including crossed-out passages and footnotes and additional notes which cannot be linked to any particular passage of the main text. For the reader to have some idea of both results in an objective way, I can say that Parsons' whole reconstruction (main text, footnotes, loose notes) has over 10,000 words, and my own 13,000.

7.5 V

Two years after having read the Gibbs Lecture, Gödel accepted the Schilpp invitation to contribute to the Carnap volume. Perhaps that may be why he decided not to publish the former lecture, for an important part of the kernel of the 1951 text contained a strong criticism of Carnap's position, and he decided to concentrate on a full development of those criticisms. From 1953 to 1959 Gödel wrote up to six versions of an essay, entitled "Is mathematics syntax of language?", which was finally not submitted. The editors of this volume offer us versions III and V, adducing that III is the richest in content and that V is the clearest version of Gödel's final position. I think this is true, but other arguments could be adduced to have chosen other versions.[8]

I think the best way to briefly describe Gödel's endeavors in the essays appearing here is by quoting from his own summary in version V. He says there that the kernel of the position to be refuted can be expressed through three assertions: "I. Mathematical intuition can be replaced by conventions about the use of symbols and their applications. II. ...there do not exist mathematical objects and facts. Mathematical propositions are void of content. III. ...the a priori validity of mathematics [is] compatible with strict empiricism." (p. 356).

On the contrary, Gödel says, "these assertions, for an adequate interpretation of the terms occurring in them (such as "content", "disprove", "replace", etc.), turn out to be wrong. Moreover, I believe, it can be shown directly that the arguments which may be adduced in favor of these assertions, including the existence of actual elaborations of the syntactical scheme, are all fallacious." (p. 357). One of the most powerful arguments used by Gödel is based on his famous metamathematical results (instead of starting with them, as in the Gibbs Lecture). Thus, in connection to assertion I, Gödel says that although it is true that we can replace mathematical intuition by conventional rules about symbols, "What must be known is that the rules, by themselves, do not imply the truth or falsehood of any proposition expressing an empirical fact. Such rules may be called 'admissible'. Admissibility, for *our* mathematics, entails consistency. For from an inconsistency *all* propositions, the empirical ones included, could be derived" (p. 357). Thus, the rules in question cannot be *really* conventional, for to prove that consistency "an intuition of the same power" is needed.

In the course of the analyses and the detailed discussions, Gödel often refers to a large number of items from the technical and philosophical literature (starting, of course, with Carnap's *Logical Syntax of Language*), which is not only mentioned in the footnotes, but sometimes even discussed in the main text. It is then not difficult to imagine the spectacular impression the essay would have caused had it been published during Gödel's lifetime. I think this expected outcome, and the consequent

[8] In my edition of Gödel's essays (see footnote 3) I chose versions II and VI. Among other things, I found version II a more finished one and thought that version VI, being the last of the series, should be published at any rate, especially if complemented with some comparative study of V, which I did in the footnotes. Fortunately, this has led to a situation in which four versions of the Carnap essay are now published, and the two versions omitted are only I, which is just the handwritten version of II, and IV, which seems to be only the beginning of an attempt of reduction conducing to V and VI.

controversy, always feared by Gödel, may have been one of the main reasons he decided finally not to submit the paper for publication.

The introductory note by Warren Goldfarb is quite useful. It provides the reader with some information on the history of the several versions and the reasons for their abandonment, together with some clear expositions of Gödel's arguments and some criticisms of them. The historico-philosophical context offered, however, seems to this writer somewhat sparse, as well as the literature related to these topics, which is only partially cited.

7.6 VI

We come now to the last philosophical text to be referred to here, the one entitled "The modern development of the foundations of mathematics in the light of philosophy". It was written about 1961 as a lecture which was never actually given. It is a very curious philosophical manuscript in which Gödel places the development of the different conceptions of the philosophy of mathematics into a twofold general frame of worldviews. He writes: "I believe that the most fruitful principle for gaining an overall view of the possible world-views will be to divide them according to the degree and the manner of their affinity to or, respectively, turning away from metaphysics (or religion). In this way we immediately obtain a division into two groups: skepticism, materialism and positivism stand on one side, spiritualism, idealism and theology on the other" (p. 375).

Later on, Gödel adds optimism and, so it seems, apriority to the second group, which seems to be the one he thinks to be correct. Thus, we have for the first time a clear scheme to more or less clearly insert his position on the foundations and philosophy of mathematics into a general view of the world. If this is so, a platonistic philosophy of mathematics should be the only one to give justice to an "optimistic", non-positivistic *Weltanschauung*. Also, the general thesis of this writer that Gödel was mainly a philosopher in search of technical results to "verify" his philosophical doctrines seems now to be more plausible.

The second part of the text is devoted to defending the position that to gain insight into the problems related to the foundations of mathematics, i.e. the meaning of the concepts and the truth of the axioms involved, the most fruitful way is not Hilbert's formalism, but something like Husserl's phenomenology: "Here clarification of meaning consists in focusing more sharply on the concepts concerned by directing our attention in a certain way, namely, onto our acts in the use of these concepts, onto our powers in carrying out our acts, etc." (p. 383). Finally, Gödel insists on his position in former manuscripts that Kant's conception of mathematics also can be useful in this respect.

Føllesdal's introductory note is devoted to summarize Gödel's ideas and to relate them to Husserl's and also with the ideas of others appearing in the rest of the manuscripts edited here, mainly in connection to realism and intuition.

As a whole this volume is as indispensable as the two former ones for any serious student of Gödel's ideas and achievements, but in this case, it is also indispensable for philosophers interested in logic and mathematics. The fourth (and last?) volume of this formidable series will be devoted to Gödel's correspondence, so we should look forward to having it to study.

Departamento de Lógica y Filosofía de la Ciencia
Universidad de Valencia

Chapter 8
Nominal Definitions and Logical Consequence in the Peano School

Abstract This paper is devoted to show the development of some of the model-theoretic ideas which are clearly present in the main members of the Peano school (Peano himself, Burali-Forti, Pieri and Padoa) as a result of their conception of nominal definitions. Also, their semantic definition of logical consequence (Pieri, Padoa) is viewed as one of the outcomes of that conception. Some examples of their use of the expression "nominal definition" are presented first. Second, the main advantages of this kind of definition, as they saw them, are briefly explained, mainly in a philosophical context. Finally, already in the kernel of the paper, some of the details of the model-theoretic view itself are shown, first in Peano, then in Pieri and Padoa, including in both cases some study of their semantic definitions of logical consequence (Former versions of parts of this paper were read in the symposium "Logics and the foundations of mathematics (1885–1905)", organized by I. Grattan-Guinness and myself, as a part of the XIXth International Congress of History of Science (Zaragoza, August, 1993), and also in the Department of Mathematics, University of Genova, June, 1995. For the preparation of this paper resources provided by the grant DGICYf PS93-0220 are gratefully acknowledged).

Keywords G. Peano · M. Pieri · A. Padoa · Logical consequence · Nominal definition

Mathematics Subject Classification Primary 03-02 · Secondary 03A05

8.1 The Expression "Nominal Definition"

The members of the Peano School used not only the expression "nominal definition", but also similar others, like "definition of name", or "symbolic definition". I think they chose the expression to avoid philosophical possible problems, given the traditional distinction between "real" and "nominal", and the danger of regarding mathematical definitions as the ones devoted to clarify the true "essence" of certain "objects" existing by themselves, in the Platonistic style. Thus, Pieri tells us something very

F. Rodríguez-Consuegra, *Russell, Gödel, Tarski*, Studies in Universal Logic,
https://doi.org/10.1007/978-3-032-08945-8_8

explicit about the distinction mentioned. He said that definitions of things, which are opposed to nominal definitions, are definitions of entities in themselves, and are constituted by a system of predicates which are sufficient to qualify a subject. Also, he wrote that the expression "nominal definition" must be preferred to exclude "real" definitions, which are definitions of things (1899, 2). However, they usually insisted that in mathematics most of the definitions are nominal in the sense that they are simply useful conventions.

Regarding examples of "nominal definitions" things are easier. As a nominal definition is the mere convention to abbreviate a group of symbols *a* with a short name *x,* so its common form is this:

$$x = a \,[\text{Def.}],$$

and so almost every definition in the writings of the members of the Peano School is a nominal definition (see Ch. 3 of my 1991). As for particular examples which they used to give to the reader when explaining their use of "nominal definition", we can for instance mention Peano's efforts to classify nominal mathematical definitions into two classes: those referred to words which cannot be found in ordinary language (e.g. prime number), and those which try to make more precise the sense of words already pertaining to ordinary language (i.e., circle) (Peano OS2, 103). A particular example actually given in Peano (OS2, 166) is this:

$$\text{Np} = (\text{N} + 1) - [(\text{N} + 1) \times (\text{N} + 1)]\text{Def.}$$

(i.e., Prime number = number which is greater than one and cannot be decomposed into two numbers greater than one.), and the following is another, this time containing an initial "hypothesis" (OS2, 208):

$$\text{a, b} \in \text{K.} \supset\!: .\text{a} \supset \text{b.} =\!: \text{x} \in \text{a.} \supset_{\text{x}} .\text{x} \in \text{b} \quad \text{Def.}$$

(In this case a "logical" symbol is defined—introduced, so it can be done only "in use". Note that apparently the symbol appears again in the *definiens,* but this is not really so, because in that case it means the conditional, together with universal quantification.)

8.2 The Advantages of Nominal Definitions

I have already said something on this when I wrote that the members of the School chose nominal definitions to avoid philosophical problems with "real" definitions. But the important point here is the role of nominal definitions as opposed, not to real ones, but to the other two main kinds of definitions they admitted: definitions by abstraction (e.g., two straight lines are parallel iff they have the same direction) and by postulates (e.g., the concept of number as "defined" by the famous five postulates).

Here the attitude was not the same for all the members of the School. Peano tended to accept the three kinds, although he always gave nominal definitions when he could, which means that he gave definitions by abstraction and by postulates only when he could not avoid them. But there was an instance of definition by abstraction (that of equality of numbers in terms of the bijective mapping between their corresponding classes) which was transformed by other members of the School (Burali-Forti by following Pieri's ideas) into a nominal one (i.e., the number of a class *a* is the class of all classes similar to *a*), and then Peano did not accept the transformation (see my 1991, p. 112). Also, he at times wrote as if definitions by postulates (e.g. the Peano postulates to characterize the concept of number) could be understood as "defining" their subject somehow.

It was Burali-Forti who best explained the main advantage of nominal definitions: they give us actual *concepts,* while definitions by abstraction and by postulates give us mere *intuitions.* That's why he added that we resort to non-nominal definitions only when we do not know how to use the corresponding nominal definitions. And this is the reason he thought that non-nominal definitions can—and must—be reduced to nominal definitions (see his 1900, 296). Another way to say the same thing is by resorting to Pieri's distinction between explicit and implicit definitions. Nominal definitions are explicit, in the sense that they eliminate, or construct, the *definiendum* in terms of the *definiens.* But implicit definitions do not eliminate the *definiendum,* so they cannot be used to build up chains of definitions which, by starting from a few primitive ideas (the ones from logic), can introduce the rest of ideas of a science. This is the most important sense according to which Russell's logicism can be said to be based on the Peano School, by mainly avoiding the several places where *Peano's Formulaire* contained "gaps", i.e., places where the chain of definitions was broken by admitting definitions by abstraction or by postulates, which forced him to admit "new" entities, which were then "irreducible" to the primitive ideas of logic (I explained that in several places of my book). This is the sense in which for Burali-Forti and Pieri nominal definitions have to be preferred.

8.3 Nominal Definitions and Logical Consequence in a Model-Theoretic Context

Some members of the Peano school, remarkably Peano himself, as well as Pieri and Padoa, developed a view of axiom systems according to which they can be regarded as formal systems which can be explained by resorting to a semantic, or model-theoretic approach, where nominal definitions are going to work as schemes which, as a matter of fact, can be seen as being independent of any standard interpretation. It is under this conception that they, rather unexpectedly, arrived at what can be regarded as a full semantic conception of logical consequence. I think they have to be acknowledged as having clearly anticipated Tarski's definition of that concept in his celebrated paper of 1936. To develop this question, I will refer separately to them.

8.3.1 Peano

When Peano posed himself the problem of the status of the definitions by postulates, for instance, the status of his famous five postulates for arithmetic, he sometimes said that those postulates can be understood as that which we can obtain by abstraction from all the possible interpretations which can satisfy the postulates (which, by the way, could also be understood as another way to transform a definition by postulates into a nominal definition; see Pieri below). In this way he was somehow convinced to make precise his well-known expression that the postulates somehow (implicitly, to be sure) "define" the primitive ideas of number, zero and successor, or the conception according to which we can say that <N, 0, +> is "the" system satisfying the five postulates, for he was also convinced that the same set of postulates could be satisfied by an infinity of different systems of entities.

The problem of nominal definitions under this rather model-theoretic viewpoint is that every definition we could build up, that is every definition whose definiens is made out of the primitive ideas, would be a definition which could hardly be depending on the "standard interpretation" of those ideas, given that the same symbolical definition would continue to be valid under a different interpretation which satisfies the postulates as well. So, the constructions actually carried out by the definitions would be mere schemes void of any intuitive content, at least until we point out some particular interpretation. That would be fine under a formal viewpoint. The problem is that for Peano the arithmetical primitives do have intuitive content, which we try to grasp through the postulates, so what every nominal definition actually does is to point out a well-known content in the definiens. Thus, for Peano the intuitive or empirical content of the primitive ideas was precisely the ultimate guarantee that when we build up a whole system, we are not constructing mere void schemes to be later fulfilled by any system of entities whatsoever, but we actually are handling real entities whose ontological status cannot be dissolved into formal schemes.

To sum up: for Peano the formal aspects of an axiom system seemed to be clear disadvantages in order to try to grasp the ultimate essence of the primitive ideas, which are given to us through a previous process of empirical induction. Thus, we can say that for him the formal character of his axiom systems was somehow opposed to the epistemological character he wanted for their primitive ideas and propositions, which were to be *entities* and *facts* previously known. As we shall see, this problem was later gradually avoided by Pieri and, mainly, by Padoa.

Even so, Peano was perfectly able to use the method of counterexamples, or interpretations (or models) to prove the relative independence of certain axioms. He used the method at least since 1889, but only in 1894 he explicitly gave a theoretic account of the method (OS3, 127):

> Si può provare l'indipendenza di alcuni postulati da altri, mediante esempi. Gli esempi per provare l'indipendenza dei postulati si ottengono attribuendo ai segni non definiti, (...), dei significati affatto qualunque; e se si trova che i segni fondamentali, in questo nuovo significato, soddisfino ad un grupo di proposizioni primitive, e non a tutte, si dedurrà che queste non sono conseguenze necessarie di quelle; ossia che il secondo gruppo di proposizioni esprimono proprietà (...) que ancora non erano espresse da quelle.

> ("It is possible to prove that some postulates are independent from others by means of examples. Those examples are obtained by attributing to the undefined symbols, (...), any meanings; and if we find that the fundamental symbols, through the new meaning, do satisfy a set of primitive propositions, but not all of them, it will be inferred that the latter propositions are not necessary consequences of the former; that is to say, that the second set of propositions express properties (...) which were not yet expressed in the first set.")

It seems to me that in this way a first implicit equivalence between independence and the negation of logical consequence appeared, in spite of Peano's lack of interest in following this development.

Also, this may have been a simple consequence of Peano's usual way of understanding logical consequence from the positive side. Thus, already in *Arithmetices Principia* (1889) he wrote (viii):

> Si propositiones *a, b* entia indeterminate continent *x, y,* ..., scilicet sunt inter ipsa entia conditiones, tunc $a \supset_{x,y} b$ significat: quaecumque sunt *x, y,* ... a propositione *a* deducitur *b.*
>
> ("If propositions *a, b* contain indeterminate entities *x, y,* ..., that is, they are conditions between these same entities, then $a \supset_{x,y} b$ means: whatever *x, y,* ... may be, from proposition *a* we can deduce proposition *b.*")

However, Peano never seemed to be interested in clarifying the semantic interpretation which could be involved in this way of describing the meaning of logical implication. Let us see how Pieri and Padoa improved on this still obscure position.

8.3.2 *Pieri*

Mario Pieri contributed with three new ideas to our topic: his method to transform definitions by abstraction into nominal ones; his conception of axiom systems as mere hypothetico-deductive systems, which made possible for the first time a true model-theoretic approach, and his method to transform definitions by postulates into nominal definitions, which was a consequence of that formal conception.

Regarding Pieri's method to transform definitions by abstraction into nominal definitions, I have already explained the two instances he offered in his publications elsewhere (see my1991, 131 ff.), but I can say here that those instances were strictly equivalent to the famous one according to which Peano and Russell later defined the number of a class, not through the equality of two numbers and the bijection between their corresponding classes, but through the class of all classes similar to the original class. With that method Pieri intended to replace the obscure method of abstraction by what he described as "true definitions".

The second, and even more important contribution was Pieri's formal view of axiom systems. According to him axiom systems consist of a set of primitive ideas without meaning and a set of postulates which are neither true nor false, and a set of theorems which are derived from them through a set of logical axioms. Thus, we cannot say anything about the primitives of a system until we provide it with an interpretation of the primitive terms, and so Pieri gave us for the first time a true scheme of the new approach to axiomatics, the one opposed to the old Euclidean view. Therefore, we do not need any epistemological, or ontological justification for the primitive ideas of a system, which are merely implicitly defined by the postulates, which do not represent facts, but mere logical relations between the primitive terms. Thus, the intuitive empirical approaches by Pasch and Peano resulted completely overcome.

From this viewpoint nominal definitions are no longer intuitive constructions which build up derivative concepts out of simple ones, but mere abstract devices, no matter the fact that we may need, to reach the theory, to keep a particular interpretation of the primitive ideas in mind. This led Pieri to a generalization of two notions which had previously appeared in geometry: (i) Gergonne's principle of duality, according to which certain demonstrations are valid for dual forms of the theorems with only changing certain primitive ideas, and (ii) Klein's method to classify the different geometries according to the different properties invariant under certain group of transformations. Regarding the principle of duality Pieri thought that it was only a particular form of a more general principle: the principle of plurality, according to which the primitive ideas of a science are indeterminate except by the postulates, so that they may be replaced by many other sets of undefined terms without affecting the proofs of the theorems. Thus, this general principle concerns not only geometry, but any other deductive theory.

Concerning Klein's method of classification, Pieri wrote that it can provide with a method to select the ideas which have to be taken as primitive. In this way, as every science is usually characterized by a certain group of transformations that cannot change the essential properties which are studied by that science, then we could take as primitive concepts those which are invariant under that fundamental group of transformations, but not under a wider group (Pieri, *Opere,* p. 257). Unfortunately, Pieri did not work out these ideas nor did he apply them to his particular investigations, but they clearly show the completely modern way in which he saw the more abstract features of axiomatic theories. Needless to say, by resorting to Klein's idea Pieri can be regarded as the first known precedent of the much later use of that idea to describe formal sciences (Tarski and others).

As a whole, that formal view of the primitive ideas and nominal definitions led Pieri even to discuss the sense according to which we could say that some sort of indetermination unavoidably affects the ultimate meaning of the undefined ideas of any deductive science. At that point it seems to me that Pieri was not yet able to draw the ultimate consequences of his own model theoretic viewpoint as clearly as Padoa did it a little later. For Pieri the plurality affecting those primitive ideas does not lead to a complete indetermination: although it is true that the mind cannot grasp the complete domain of those ideas, however, the primitive propositions always allow us to decide whether or not any given object belongs to the domain of the primitive concepts. Thus, apparently Pieri did not see the possibility of non-standard interpretations of his systems, in spite of the fact that he, at least since 1906, was in possession of the concept of categoricity (*Opere,* p. 432), doubtless taken from Huntington and Veblen. Again unfortunately, Pieri did not apply this concept to any of his axiomatizations.

A further consequence of Pieri's abstract conception of definitions is that it made a true model theoretic description of implicit definitions possible. Thus, Pieri said, not only that the postulates somehow "define" the primitive terms of a theory, but also that the whole set of postulates define a global concept. For instance, in speaking of the postulates of his axiomatization of projective geometry of 1898, Pieri wrote that they define the global concept of geometric space, which can be understood as the class of all possible interpretations which are capable of satisfying the postulates *(Opere,* 155–6). Thus, he not only advanced the approach used by Russell in his *Principles of mathematics* (precisely in the sections on Geometry, located in the last part of the work), but showed a mastering of the abstract axiomatic method which has been said to be superior to Hilbert's classic approach of 1899, which is many times associated rather to the implicit definition approach, which was already criticized by Frege, and corrected by Bernays in his famous review about forty years later.

The model-theoretic approach made also possible for Pieri to apply systematically metatheoretic devices at least from 1897 on. Thus, in that year he devoted a whole appendix to his axiomatization of projective geometry in which he constructed models to demonstrate the ordinal independence of the postulates, and offered a whole model to demonstrate their consistency, and all that about two years before Hilbert's *Grundlagen.* It is true that for Pieri the only way to prove consistency was the exhibition of a model, as his later contributions to the problem show, so he explicitly supported Frege's view that the consistency of a concept means essentially the existence of some object falling under it. Curiously, when Hilbert discussed with Frege about this problem in the celebrated correspondence, he defended the view that mere consistency is enough for mathematical existence, in spite of the fact that by those years Hilbert himself had only available the method of the exhibition of a model to prove relative consistency, as his *Grundlagen* shows.

It is in the framework of this full model-theoretic context that Pieri's definition of logical consequence appeared already in 1897 (*Opere,* 60, 45, 9). There he defined independence in the usual way:

> (...) le proposizioni *P, Q, R,* ... si diranno "*indipendenti* le une dalle altre" se avvien che nessuna sia conseguenza delle rimanenti (dunque se avviene, che per ciascuna si possan trovare degli *x, y, z,* ... che non la verificano, mentre rendon soddisfatte le altre).
>
> ("(...) the propositions *P, Q, R,* ... will be said to be '*independent* from each other' if none of them is a consequence of the rest (therefore, if for any of them we can find certain *x, y, z,* ... which do not verify it, while they do satisfy the rest).")

But he had insisted before that this conception of independence was based on a clear semantic conception of the negation of logical consequence in terms of sets of objects or models, that is in clear semantic terms:

> Date (...) proposizioni condicionali *P*(*x, y, z,* ...), *Q*(*x, y, z,* ...), *R*(*x, y, z,* ...), ecc., sugli enti variabili x, y, z, ... non può cader dubbio sul valore delle asserzioni "dalle *P* e *Q non si deduce* la *R*", "la *R non è consequenza* delle *P* e *Q*"; (...) Ambo i modi null'altro esprimono che questa prosizione particolare: "*esistono* degli *x, y, z,* ... per cui son vere la P e la Q, ma non è vera la R".
>
> ("Given (...) conditional propositions *P(x, y, z,* ...), *Q*(*x, y, z,* ...), *R(x, y, z,* ...), etc., over variable entities x, y, z, ... there could not be any doubt on the meaning of the assertions '*R*

cannot be deduced from *P* and *Q*', '*R is not a consequence* of *P* and *Q*'; (...) Both express but this particular proposition: '*there are* certain *x, y, z,* ... from which P and Q are true, but R is not true'.")

Again, this semantic conception of the negation of logical consequence was said to be based on a previously introduced positive conception of the same concept:

> (...) il giudizio "da *P(a, b,* c, ...) si deduce *Q(a, b,* c, ...)" - (...) - è da retenersi il medesimo che: "qualunque siano *a, b,* c, ..., se per essi è vera *P(a, b,* c, ...) sarà altres vera *Q(a, b,* c, ...)...".
>
> ("(...) the judgement 'from *P(a, b,* c, ...) is deduced *Q(a, b,* c, ...)' - (...) - has to be kept in mind as being the same as 'for every *a, b,* c, .. , if *P(a, b,* c, ...) is true for them, *Q(a, b,* c, ...) will be true as well...")

Or, even in clearer terms:

> Se *P(x, y, z,* ...), *Q(x, y, z,* ...) sono proposizioni negli enti variabili *x, y, z,* ..., la scrittura "*P(x, y, z,* ...) $\supset_{x,y}$ *Q(x, y, z,* ...)" significa "qualunque siano *x, y,* purchè soddisfacenti a *P(x, y, z,* ...), dovranno altres verificare la *Q(x, y, z,* ...)"; e può leggersi: da *P* si deduce, rispetto ad *x, y, Q*".
>
> ("If *P(x, y, z,* ...), *Q(x, y, z,* ...) are propositions over the variable entities *x, y, z,* ..., the writing '*P(x, y, z,* ...) $\supset_{x,y}$ *Q(x, y, z,* ...)' means 'for every *x, y,* provided they satisfy *P(x, y, z,* ...), they must verify *Q(x, y, z,* ...) as well'; and so it can be read: Q is deduced from P, with regard to *x, y,* Q'.")

It is true that here Pieri did not speak explicitly about models or interpretations, and, in particular, he did not speak about "every interpretation" (or "all models"), so making the quantifiers range over every domain. It is also true that it is possible that he was simply thinking of proofs in the usual truth-functional (conditional) way, as Peano himself did before. But I am convinced he was so familiar with the model-theoretic approach that in speaking this way he had in mind something very similar to the now usual way of defining logical consequence. At any rate, only three years later Alessandro Padoa reached the full contemporary definition of logical consequence, as we are going to see to finish.

8.3.3 *Padoa*

In his "Logical introduction to any deductive theory", a part of the paper read before the 1900 Paris International Congress of Philosophy, Padoa established his general approach to his formal and model-theoretic conception of axiom systems. Today it is believed that his main contribution in that paper was his method to prove whether the undefined symbols of a system are irreducible with respect to the system of its unproved propositions, which was a further application of Peano's method to prove the independence of the postulates. For Padoa, to do that, it suffices to find, for each undefined symbol, an interpretation of the whole set of primitive terms which verifies the postulates even when the meaning of the symbol considered is suitably changed. But the interesting thing here is pointing out that the introduction of that method was only possible thanks to a general view of axiom systems as mere formal, uninterpreted schemes.

Thus, Padoa needed, before discovering his famous method, to see deductive theories in a new way, according to which the ideas and facts which usually were given as the ones represented by the undefined symbols and the unproved propositions, were actually mere empirical or psychological aids which we use in constructing the theory, but which we must regard as quite independent of the theory itself, regarded as a formal system. Thus Padoa, instead of accepting Pieri's distinction between axioms and postulates, interpreted the postulates of a formal theory as mere conditions imposed upon the undefined symbols and definitely forgot about any criterion of simplicity for them. Under this general view, he described for the first time nominal definitions as mere relations between a symbol and the symbols previously introduced which is able to single out a meaning for that symbol as soon as we choose an interpretation of the system as a whole.

However, both aspects of his completely model-theoretic conception of theories, that is, the already quite formal conception and the consequent character of nominal definitions, were exposed in a much clearer way in an unpublished manuscript of the same year entitled "Riassunto delle Conferenze su l'Algebra e la Geometria quali teorie deduttive" in which a series of lectures in the University of Rome were transcribed.[1] In addition, a clearer conception of the relationship between the usual way of proving the independence of the postulates and the notion of logical consequence can also be found in the manuscript, in which we can find even a literally Tarskian definition of logical consequence. Let us see these three points through the corresponding passages.

Regarding the formal view of axiom systems, Padoa writes:

> Noi consideremo le teorie deduttive sotto *l'aspetto formale*; immagineremo cioe che, all'inizio della teoria, I simboli che rappresentano le idee assunte quali primitive, sieno sprovisti di significato.
>
> ("We consider deductive theories under the *formal aspect*; that is, we imagine that, at the beginning of the theory, the symbols which represent the ideas assumed as primitive are devoid of meaning".)

To my knowledge this is the first use of the expression"formal" in that context in the Peano school. Concerning the role of nominal definitions under the model-theoretic conception, we can read:

> Nelle teorie deductive considerate sotto l'aspetto formate, nemmeno le Df che I logici dicono "*nominali*" individuano il significato delle *idee derivate*. Ogni Df esprime soltanto una *relazione* fra il *significato della idea definita* e *l'intepretazione delle idee primitive;* per modo che, fissata questa, quello risulta implicitamente individuato.

[1] The ms was sent to me for study by I. Grattan-Guinness, who found it in the Library of Pavia, Italy. The reader may easily imagine my astonishment in discovering Padoa's definition. We do not know yet whether those lectures were given before of after the Paris Congress of 1900. On the one hand Padoa may have given them before the Congress, whose mentioned contribution must have contained only a set of partial results; on the other, some of the details of the manuscripts might be interpreted as the result of further thinking, as for instances it takes place with his more formal general view. [See Bibliography below.]

("In the deductive theories regarded under the formal aspect the definitions which are called 'nominal' by the logicians do not single out the meaning of the *derivative ideas.* Every definition expresses only a *relationship* between the *meaning of the defined ideas* and the *interpretations of the primitive ideas;* in a way that once we set the later, the first results implicitly singled out.")

Thus, we can say that only Padoa made quite explicit the tendency towards expressing nominal definitions as mere abstract relations, which can be discovered only in an implicit way in Peano and Pieri. And we can add that this was possible thanks to Padoa's explicit effort to distinguish the logical, the semantical and the epistemological elements which were involved in the transition from the old Euclidean axiomatics to the new modern formal one.

Finally, let us see Padoa's model-theoretic definition of logical consequence. Already in his contribution to the Paris Philosophy Congress Padoa, by following the tradition of Peano and Pieri, wrote that in order to prove the independence of an unproved proposition it suffices with exhibiting an interpretation under which the given proposition could be false and all the others true. However, Padoa also made explicit the ultimate reason for that, which the rest of authors of the school left implicit: when we establish an interpretation of the undefined symbols that verifies the unproved propositions but one, "then this proposition is not a logical consequence of the other propositions; that is, it is not possible to deduce the proposition in question from the other unproved propositions".

In the unpublished manuscript I referred before the corresponding passage contained a full definition of what we call today "logical consequence" on these lines, which precedes the introduction of the method to proving independence:

> Se una Pp, ad es. α, fosse deducibili dalle precedenti, ciò significherebbe che ogni interpretazione delle idee primitive, la quale verifichi le Pp precedenti la Pα, deve pure verificare la Pα.
>
> ("If a primitive proposition, say α, was deducible from the precedent propositions, that would mean that every interpretation of the primitive ideas which verify those primitive propositions must also verify the proposition Pα".)

This is to me strictly equivalent to Tarski's famous definition, according to which a given proposition φ is a logical consequence of a set Γ of propositions iff every model which satisfies Γ also satisfies φ.[2] Therefore, only Padoa was able to make completely explicit the fact that the old method to prove independence was implicitly based on an underlying concept of logical consequence in the modern sense. (No matter the fact that he did not say anything about the difference between derivability and logical consequence, that is, the difference, which was discovered much later, between syntactic and semantic deducibility.)

The fact that only Padoa was able to do that seems to me to be due to his complete mastery of the formal, model-theoretic approach. Before Padoa, every member of

---

[2] In a recent trip to Italy I have discovered that Padoa (and also Burali-Forti) used similar semantic definitions of logical consequence in some publications, now unfortunately forgotten. I hope to be able to write a further paper on this particular subject in order to develop some of the ideas appearing here with the help of those additional definitions. I have recently dealt with Pieri's and Padoa's ideas in my papers 1996a, b and c.

the school knew about the usual methods of proving consistency and independence, which were essentially introduced by the discoverers of non-Euclidean geometries, which showed that the postulate of the parallels was independent of the rest of Euclidean postulates by finding the corresponding interpretations, and also gave interpretations of the new geometries which showed that they were consistent if Euclidean geometry was consistent (Beltrami, Klein). But only Padoa, who, in addition, usually saw axiom systems under the model-theoretic viewpoint, and was especially interested in analyzing the abstract structure of deductive theories as a goal which was interesting in itself, was able to draw the corresponding conclusions.

To finish, I can try to make quite explicit Padoa's argument relating independence and logical consequence (for: Γ = set of statements; $\varphi \in \Gamma$; M = model; sat = satisfies; ind = is independent of). The usual definition of independence in model-theoretic terms, assuming the consistency of the whole system, is:

$$\varphi \text{ ind } \Gamma \leftrightarrow \exists \mathrm{M}\ (\mathrm{M} \text{ sat } \Gamma \wedge \mathrm{M} \text{ sat} \neg \varphi)$$

by replacing 'ind' by '$\neg \models$', and '$\exists \ldots \neg$' by '$\neg\forall$', we obtain

$$\Gamma \neg \models \varphi \leftrightarrow \forall \mathrm{M}\ (\mathrm{M} \text{ sat } \Gamma \rightarrow \mathrm{M} \text{ sat } \varphi)$$

therefore, logical consequence can be defined this way:

$$\Gamma \models \varphi \leftrightarrow \forall \mathrm{M}\ (\mathrm{M} \text{ sat } \Gamma \rightarrow \mathrm{M} \text{ sat } \varphi)$$

which is, precisely, what Padoa, and much later Tarski, actually did.

References

Burali-Forti: 1900, 'Surles differentes methodes logiques pour la definition du nombre reel', *Congres Int. de Phil.,* Paris, 1900, Paris, Colin, vol. III, 289–307.

Padoa, A: 1900a, 'Essai d'une theorie algebrique des nombres entiers, precede d'une introduction logique a une theorie deductive quelconque', *Congres Int. de Phil.,* Paris, 1900, Paris, Colin, vol. III, 309–65. Trad. ing. de la primera parte en van Heijenoort (1967), 118–23.

Padoa, A: 1900b, 'Riassunto delle Conferenze su l'Algebra e la Geometria quali teorie deduttive', Sort of unpublished manuscript extant in the Pavia Library, discovered by I, Grattan-Guinness, who sent it to me for study. I was astonished to discover that definition. [There are more details about Padoa's ms in my book with E. A. Marchisotto and J. T. Smith, *The Legacy of Mario Pieri in Foundations and Philosophy of Mathematics.* New York and Basel, Birkhäuser, 2021.]

Peano, G.: 1957, *Opere scelte* (OS), 3 vols., U. Cassina (ed.), Roma, Cremonese, 1957–9.

Peano, G.: 1889, *Arithmetices principia, nova methodo exposita.* Rep. and Spanish translation by J. Velarde, Oviedo, Pentalfa, 1979.

Peano, G.: 1894, 'Sui fondamenti della geometria', OS3, 115–57.

Pieri, M.: 1980, *Opere sui fondamenti delta matematica,* Roma, Cremonese.

Peano, G.: 1898, 'I principii della geometria di posizione composti in sistema logico deduttivo', *Memorie della Reale Accademia delle Scienze di Torino* 48, 1–62. Rep. en *Opere.*

Peano, G.: 1899, 'Della geometria elementare come sistema ipotetico deduttivo', *Memorie delta Reale Accademia delle Scienze di Torino* 49, 173–222. Rep. en *Opere.*

Rodriguez-Consuegra, F.: 1991, *The mathematical philosophy of Bertrand Russell: origins and development,* Basilea, Boston y Berlin, Birkhauser. 2nd print. in 1993.

Rodriguez-Consuegra, F.: 1993a, (with EA. Marchisotto), 'The work of Mario Pieri in the philosophy and foundations of mathematics', *Hist. Phil. Logic* 14, 215–20.

Rodriguez-Consuegra, F.: 1993b, 'Mario Pieri y la naturaleza de la axiomatica', in Bustos, E., Echeverria,]., Perez, E. and Sanchez, M.I. (eds.): *Actas del I Congreso de la Sdad de L6gica, Metod. y Filas. de la Ciencia en España,* Madrid, UNED, 1993, 502–505.

Rodriguez-Consuegra, F.: 1996a, La definicion semantica de consecuencia 16gica: descubrimiento de la escuela de Peano', *Arbor* 1996, in print. [CLIV, 606 (1996): 103–116.]

Rodriguez-Consuegra, F.: 1996b, 'Mario Pieri y la filosoffa de las ciencias deductivas', in F. Zalamea (ed.): *La trenza y el velo. Lazos entre historia, matemáticas, lógica y filosofía,* Universidad Nacional de Colombia, Bogotá, 1996, in print.

Rodriguez-Consuegra, F.: 1996c, 'La vía negativa hacia el concepto de consecuencia 16gica', in the same as 1996b. [This and the former paper were translated into English, expanded and much improved, then published in E. A. Marchisotto and J. T. Smith, *The Legacy of Mario Pieri in Foundations and Philosophy of Mathematics.* New York and Basel, Birkhäuser, 2021.]

Departamento de Lógica y Filosofía de la ciencia
Universidad de Valencia

Chapter 9
Philosophy in Hao Wang's Conversations with Gödel

Abstract This paper is a critical Essay Review of Hao Wang, *A logical Journey. From Gödel to Philosophy.* Cambridge and London: The MIT Press, 1996, xiv + 391. This is a long-awaited book. It contains Hao Wangs records of his conversations with Kurt Gödel (1906–1978) from 1971 to 1976, ordered by subjects, and some reflections by Wang about Gödel's life, personality and ideas. The contents of the conversations are mainly philosophical in nature, which is very understandable; although Gödel almost never dared to make his philosophical ideas public, he sometimes felt the need of some philosophical interchange. Consequently, I will consider only the philosophical side of the book, though some other interesting things about Gödel can be found here.

Keywords K. Gödel · Logic · Mathematics · Philosophy

Mathematics Subject Classification Primary 01A70 · Secondary 00A30

This essay review is divided into seven sections The first and second of these are devoted to the origins of the book, as it was actually published, and to the origins of the conversations themselves. The third tries to present a global scheme of Gödel's more general philosophical conceptions. Sections 9.6–9.7 will sum up the main ideas of what he thought about philosophy of mind, of mathematics and of logic and set theory. Finally, I will attempt some assessment of the book itself, and also of some of the philosophical ideas appearing here for the first time, in the context of Gödel's well-known ideas available in his previous publications, including the recent ones containing his philosophical unpublished (Gödel 1995a, 1995b).[1]

[1] The first of Gödel's philosophical manuscripts to be published appeared in my "Un inédito de Gödel contra el convencionalismo: historia, análisis, contexto y traducción," *Arbor* CXLII, 558-559-560, 1992: pp. 323–348.

F. Rodríguez-Consuegra, *Russell, Gödel, Tarski*, Studies in Universal Logic,
https://doi.org/10.1007/978-3-032-08945-8_9

9.1 The Origins of the Book

For many years Wang did not find a satisfactory way to publish the records of his conversations with Gödel. He told me personally in the early nineties that after many different attempts, he had arrived at a general presentation where the different fragments appeared in chronological order. Yet he was not satisfied, mainly because the global result was too sketchy, and many fragments were rather loose in the general context. After that, in correspondence, he told me that he was trying to find someone to assist him for this matter, and even sent me a complete typed copy, from which I quoted some fragments concerning some of Gödel's philosophical views (see Gödel 1995b, pp. 27, 33).

Hao Wang finally proposed Michael Detlefsen to assist him in completing the projected book, but after taking some time to consider the offer, Detlefsen declined and suggested a former student of him in Notre Dame, Anthony Everett, for the job, who was accepted by Wang. Nevertheless, Detlefsen continued to be in touch with Wang, both in correspondence and over the phone, about his writings, some of which he read and made comments about. Yet Wang must have been somehow annoyed with Detlefsen's former negative decision, for he is not mentioned in the published book.

Wang does mention Everett in the Preface (p. xiii), to credit him for his help during two years in making an index and an outline of the notes, an index of the book, and for his comments on a previous draft of the book. Everett is at present happy with that acknowledgement, but it seems to me that it is not as generous as it should have been, according to Everetts own description of his actual work on the book, which I now quote because of its importance in understanding the whole history:

> Wang had several hundred pages of fragmentary notes of his conversations with Godel [sic], interspersed with some of Wang's own observations. For many years he had been trying to work these into a book of some form, a final tribute to Godel [*sic*]. I dont think he was at all sure how best to do this. I sorted through these notes and indexed them. I also tried to order them into various topics, order the notes within each topic in a philosophically coherent manner, and order the topics. So, the final order in which the notes appear in the book is probably largely my responsibility.
>
> After talking with Wang, I understood my task to be this: to try to *re-construct* or *re-write* Godel's [*sic*] observations into a more coherent and flowing body of text that could stand on its own. I was not completely sure how much leeway I was allowed. I am pretty sure that Wang thought this was my task too, but one can never be completely certain. As I saw it the result would not have quite been Godel [*sic*], but in Godel's [*sic*] spirit, perhaps in much the same way that people try to complete the novels and symphonies of dead authors and composers. It would not have been presented as "Godel's word" [*sic*] but as an explicit reconstruction of it.
>
> Anyway, I spent much time trying to do this. Obviously, this was a difficult way of proceeding, and maybe the result would have been of only limited interest. Quite rightly, I think, Wang disliked the result of my work and resolved to include the fragments as they stood. He wrote connecting text and comments quite quickly I believe. I had read and commented upon drafts of most, but not all, of the chapters before Wang died. To some extent the form of the book may still be partly a compromise, since Wang was very ill and knew that he was dying, and wanted to see the thing finished while he lived.

> The original idea as suggested by Mic [Detlefsen], and as I understood it, was that the result should have been co-edited by Wang and me. However, since Wang ended up writing most of the book it is really his own child and I cannot take much credit. I see my own most important contribution to the book as so horrifying Wang that they allowed him to better understand what he wanted the book to be and what he didnt want it to be. Perhaps they also helped motivate him to finish it (out of fear that my "textual reconstruction" would otherwise be published?).

Let us finish this section with a comparison between the final presentation of the notes with the original one in the original manuscript of the book which I—and also Everett—mentioned above. In what seems to be its last stage, that manuscript was entitled *For perspicuous objectivity. Discussions with Gödel and Wittgenstein.* It was divided into two parts. The first one constitutes a book in itself, containing five chapters, respectively entitled: (1) "To philosophy, from the views of Wittgenstein and Gödel on mathematics"; (2) "Five islands of agreement with Gödel"; (3) "Creation and discovery: on mathematical realism"; (4) "Dialectic of intuition and idealization: on conceptual realism", and (5) "Objective certainty and the nature of mathematics". Be-sides, there are a preface, a section of abbreviations and reference, an introduction, and several pages of notes after each chapter. As a whole, the book—which is dedicated "To the memory of Kurt Gödel"—extends to xiv + 178 typed pages and seems to be ready for publication. I think it should be published as it is.

The second part of the manuscript, entitled "The conversations", starts with an introduction (17 pp.), and is divided into four chapters: 1. "October 1971 to February 1972" (77 pp.); 2. "March to December 1972" (46 pp.); 3. "October 1975 to January 1976" (79 pp.) and 4. "February to June 1976" (40 pp.). Probably Wang thought at one time of publishing an entire book including all this material and then found the idea rather unconvincing.

Some of the things Wang says in the manuscripts do appear in the book under review, here and there. Unfortunately, the numbering system of the Gödel passages has been heavily modified. In the published book they appear simply with three numbers, the first being the number of the chapter, the second the number of the section, and the third the number of the particular passage (e.g., 5.3.22); also, the remaining quotations, whether by Gödel—in published or unpublished writings—or by other authors, follow the same pattern. On the contrary, the system of the manuscript was described by Wang this way: "I have split the conversations into about 520 segments, which are numbered according to their place in the year and the month. For instance, 602.1 refers to the first segment (1) in February (02) of 1976 (6)". Obviously enough, the original system would have been preferable, as it would give the reader and the scholar additional information about the particular time in which the idea was originally formulated, which is necessary to understand Gödel's preferences according to particular stages of the evolution of his ideas. In the actual book the fragments appear divided according to the subject matter dealt with in each of them; yet this could have been done in the same way but preserving the old numbering system.

9.2 The Origins and Context of the Conversations

After some more or less casual contacts, the relation between Wang and Gödel began in 1967, when Wang sent Gödel a draft of a manuscript on Skolem for comments, and Gödel provided some. A few paragraphs of Gödel's letters were finally published in Wangs *From Mathematics to Philosophy* (Wang 1974) and are well-known. As it happened, Gödel invited Wang in January 1970 to visit him, and then Wang asked Gödel to comment on a draft of his projected book, which Gödel must have read carefully, especially the parts where his name occurred. As a matter of fact, the conversations began by making extensive comments on those parts, then they extended to other parts.

From October 1971 to December 1972 Wang and Gödel met regularly in Gödel's office in Princeton, from 11 am to 1 or 1:30 pm. They had twenty sessions where Wang was allowed to take notes. Unfortunately, he did not bring a tape recorder ("at that time I did not find the idea congenial and did not propose it", p. 136). Usually, Gödel brought slips of paper with the ideas he wished to talk about and Wang took notes during the sessions. Yet he recognized that the final notes remained in a very unsatisfactory state, despite his efforts to reorganize them after each session.

They usually started talking about philosophy and Wangs manuscript. After it was sent to the printer in June 1972, Gödel's philosophical views constituted the nutshell of the conversations. Several times Gödel wanted to show Wang some of his unpublished manuscripts—probably the Gibbs lecture, given in 1951, in the first instance—but he always wound up by finding the manuscript "not yet in a form fit to be shown" (p. 136). Also, Gödel declined to expose his philosophy in a systematic way, arguing that it was not yet developed to the point to lecture on it.

In 1975 Wang became a visitor at the Institute for Advanced Study for more than one year, and they resumed the contacts. Yet Gödel had stopped going to his office—by then he had health problems, so most of the conversations were over the phone, and based upon written versions of former discussions which Wang had previously sent him.

Wangs whole conception of his reports is important. He says that some of them were fully approved by Gödel, while others were not. Some were never seen by Gödel. As a result, Wang tells us that Gödel "would not have wished to publish much of the material in the form I quote, and it is quite possible that there are places where I am mistaken about what he actually says" (p. 137). Also, Wang adds that he made no serious attempt to look to Gödel's *Nachlass*, whose materials might prove some of his reports to be mistaken. The reader is warned: "what I report in this book must be understood *cum grano salis*" (p. 137).

9.3 Gödel's Global Philosophical Viewpoint

Gödel's own philosophy took a clear side in spiritualism, rationalism, idealism and theology, against skepticism, materialism and positivism. His philosophical global viewpoint was summarized by him in a manuscript from 1960 this way:

> The world is rational. Human reason can, in principle, be developed more highly. There are systematic methods for the solution of all problems. There are other worlds and rational beings of a different and higher kind. The world in which we live is not the only one in which we shall live or have lived. There is incomparably more knowable *a priori* than is currently known. The development of human thought since the Renaissance is thoroughly intelligible. Reason in mankind will be developed in every direction. Formal rights comprise a real science. Materialism is false. The higher beings are connected to the others by analogy, not by composition. Concepts have an objective existence. There is a scientific philosophy and theology, which deals with concepts of the highest abstractness; and this is also most highly fruitful for science. Religions are, for the most part, bad – but religion is not. (p. 316)

Accordingly, Gödel defended a dualistic ontology based on a sort of Leibnizian monadology, an epistemology based on a probable, rationalistic knowledge, which should be always guided by observation—including intuition—and a series of opinions about the philosophers—either from the past or contemporaries—which were inspired by them.

His conception of the monads is highly speculative. Monads, which in the end should compose matter, are neither material nor spatial, although they act in space. Also, they have a consciousness and maintain relationships with other particles, while having "something inside". Their spiritual nature will be correctly understood just when "the true theory of physics is found" (p. 292). Among the different kinds of being, concepts and objects seem to be the essential ones, although the relations between them and monads are by no means clear. Concepts are wholes, composed of primitive concepts ("such as negation, existence, conjunction, universality, object, concept, whole, meaning, and so on", p. 295), and are wholes of an organic nature, in a way that sets are not. As for objects, including monads and sets, they can be of different kinds, including physical and mathematical ones. The following quotation may perhaps clarify this a little:

> A set is a special kind of whole. Sets are unities which are just the multitude; but generally wholes are more than multitudes which are also unities. That is why sets are a limiting case of wholes. A whole must have parts. A monad is a unity but not a whole because it is indivisible: it is only an *uneigentlich* [improper] whole. Primitive concepts, like monads, are unities which are not wholes, because they are not composed of parts. (p. 296)

Be that as it may, primitive concepts are to be looked for by us, and then systematically exposed to give an account of the world; thus, we come to knowledge.

All knowledge depends upon observation of the different kinds of beings; it is not infallible, but every mistake is due to the empirical component. It can be corrected through our whole experience, properly guided by pure reason, which "does not commit mistakes" by itself and seems to include some "probable intuition" (p. 291). Thus, skepticism is to be rejected. Yet "experience" should not be understood in the empirical manner, for the sort of observation presupposed here includes *Wesenschau*, which is explained by Gödel as "essential intuition, grasp of essence, categorial intuition, perception of concepts" (p. 292). Therefore, what is usually called experience is irrelevant. On the contrary, everything is inside ourselves: "Dont collect data. If you know everything about yourself, you know everything. There is no use in burdening yourself with a lot of data. Once you understand yourself, you understand human nature and then the rest follows." The results of the investigations ruled by the correct use of our knowledge should be expressed as a sort of grand axiomatic, theoretical metaphysics, apparently based on some use of mathematical logic, somehow conceived as the proper science of general concepts.

The relation between axiomatics and ontology is clearly expressed when Gödel writes: "The axioms correspond to the concepts, and the models which satisfy them correspond to the objects. The representations give the relation between concepts and objects" (p. 141). Also, the relation between ontology and epistemology can be better understood through this: "Sets are objects but concepts are not objects. We perceive objects and understand concepts. Understanding is a different kind of perception: it is a step in the direction of reduction to the last cause" (p. 235).

Upon these bases it is not very difficult to figure out Gödel's philosophical friends and also his enemies, which constitutes the kernel of chapter 5 of the book. Among the first Plato, Kant and Husserl have to be mentioned. We can try to summarize his views about them. Plato can be regarded as the founder of philosophy because he studied the definition of concepts. Kant is to be acknowledged because, although his philosophy somehow gives the pre-eminence to architectonic over an absolute and systematic starting-point, however he looked for a transcendental—yet rather subjective—philosophy where "all categories should be reduced to something more fundamental" (p. 171). Husserl should be central in what Gödel thought to be the true method of philosophy, because his phenomenology tries to go back to the absolute foundations of our knowledge, "to the process of how we form the knowledge, and to uncovering what is given to us from inside" (p. 167). Yet there are numerous shades in the fragments which have to be carefully analyzed to obtain a more articulate viewpoint about them.

As for Gödel's philosophical enemies, most of his critical passages are written against the logical positivists, and other "linguistic philosophers", Carnap and Wittgenstein being the names appearing more often. However, as the kernel of his criticisms are now well-known, after other publications by Wang, and also through the recent appearance of some of Gödel's unpublished manuscripts, I will leave those passages without further comment.

9.4 Philosophy of Mind: Spiritualism

Gödel's main views in the field of the philosophy of mind seem to be a consequence of his struggle against materialism and positivism, and can be summarized by saying that for him the mind cannot work like a computer, and that the brain is not enough to explain mental phenomena, so even if the brain were shown to be like a computer, very interesting mental abilities have yet to be explained. From the publication of Wangs *From mathematics to philosophy* in 1974 we know Gödel's main arguments against any possible attempt to identify minds and computers (in the context of his comments about Turing machines). They were: that the use of the mind is in constant development, while computers are static; that the states of every computer are necessarily finite, while mental states might converge to infinity because of their development, and that there may exist mental procedures which are not mechanical in nature (he was obviously thinking of mathematical intuition, which for Gödel cannot be reduced to any mechanical procedure).

Also, from his Gibbs lecture, given in 1951 but published for the first time in 1994 (in Spanish) and 1995 (in English) (see the works cited in Gödel 1995b), we know that for Gödel his celebrated incompleteness results show that mathematics is inexhaustible, so that the human mind is unable to mechanize all its mathematical intuitions. This was expressed by him as an interesting logical disjunction: either the system of all demonstrable propositions (subjective mathematics) surpasses all machines, or the system of all true mathematical propositions (objective mathematics) surpasses subjective mathematics. If the first, the human mind cannot be reduced to the brain; if the second, mathematical objects and facts cannot be our creation, so they are independent of our mental acts. Obviously enough, both alternatives were to be unacceptable for materialists, so they together may be indirectly argued in favor of spiritualism.

Most of the passages published in Wangs new book concerning Gödel's philosophy of mind express very similar ideas, or contain comments on them, as for instance it is the case with these:

> The incompleteness results do not rule out the possibility that there is a theorem-proving computer which is in fact equivalent to mathematical intuition. But they imply that, in such a – highly unlikely for other reasons – case, either we do not know the exact specification of the computer or we do not know that it works correctly. My incompleteness theorem makes it likely that mind is not mechanical, or else mind cannot understand its own mechanism. If my result is taken together with the rationalistic attitude which Hilbert had and which was not refuted by my results, then [we can infer] the sharp result that the mind is not mechanical. This is so, because, if the mind were a machine, there should, contrary to this rationalistic attitude, exist number-theoretic questions undecidable for the human mind. (pp. 186–187)

But there also are other passages in which spiritualism is openly defended: "Even if the finite brain cannot store an infinite amount of information, the spirit may be able to. The brain is a computing machine connected with a spirit" (p. 193). As usual then, Gödel was much more cautious in his publishable writings than in his private reflections.

For similar general reasons, Gödel was also against psychophysical parallelism: it is not compatible with his strong conviction that the mind cannot be reduced to the brain. But in this case a new argument appears here, which connects the issue with empirical matters: parallelism can be disproved scientifically, as "there arent enough nerve cells to perform the observable operations of the mind" (p. 190). Similarly, Gödel wrote that the brain cannot be explained in the usual Darwinian way, as "life force" works according to laws which are neither simple, nor mechanical. Yet the most specific argument in the same direction seems to be just that the formation of a human body by the laws of physics "is as unlikely as the separation by chance of the atmosphere into its components" (p. 192).

9.5 Philosophy of Mathematics: Objectivism

Objectivism, or Platonism, was the kernel of Gödel's philosophy of mathematics, and consisted of the belief that mathematical objects and concepts (the entities referred to by mathematical symbols) and facts (those expressed by mathematical propositions) are not our creation; yet they do objectively exist with total independence from the existence and working of our minds. This philosophical position was fully developed in the Gibbs lecture and in Gödel's essay on Carnap.

The Gibbs lecture studied the philosophical implications of the incompleteness theorems. According to it, mathematics is inexhaustible, so we cannot do mathematics without intuition, which cannot be replaced by algorithmical methods. Mathematical concepts and facts are objective, and the human mind can perceive them in a way which cannot be done by any finite machine. As for the Carnap essay, mathematics cannot be reduced to a formal system of logical syntax of language, for—after the incompleteness results—no similar system could do the work unless it makes use of concepts similarly powerful to the ones to be reduced, so any attempt on these lines would be vacuous. However, these writings by Gödel are already published, so I will limit myself here to identifying some of the ideas which are not appearing in them. In the conversations with Wang, Gödel gave similar arguments, but they were more simply stated and emphasized the importance of number theory. He clearly said that the "real argument" for objectivism is simply that, as we know that many propositions about natural numbers are true, and as we believe that many related conjectures (e.g., Goldbach's) make sense, then "there must be objective facts about natural numbers" and these facts "must refer to objects" which "are unchangeable in time" (p. 211). As for the argument against conventionalism (i.e., against Carnap), Gödel added that logic and mathematics must have a real content which can be seen by studying number theory, where "we encounter facts that are independent of arbitrary conventions. These facts must have a content because the consistency of number theory cannot be based on trivial facts, since it is not even known in the strong sense of knowing" (p. 212).

However, Gödel was aware that as long as we go up to higher mathematics, the solid ground seems somehow to vanish. In the published writings he used to talk about the help of intuition for that case, but here he used the rather loose notion of "idealization":

> Strictly speaking we only have clear propositions about physically given sets and then only about simple examples of them. If you give up idealization, then mathematics disappears. Consequently it is subjective matter where you want to stop on the ladder of idealization... Without idealizations nothing remains: there would be no mathematics at all, except the

part about small numbers. It is arbitrary to stop anywhere along the path of more and more idealizations. We move from intuitionistic to classical mathematics and then to set theory, with decreasing certainty. The increasing degree of uncertainty begins [at the region] between classical mathematics and set theory. Only as mathematics is developed more and more, the overall certainty goes up. The relative degrees remain the same. (p. 217)

It seems that for Gödel idealization was not only a stronger notion than intuition, but something somehow related to his holistic overall conception of mathematics and general science; also, it might be connected to Kant's "regulative" use of some ideal concepts in order to obtain progress in our research, which is always inspired by the principle of getting more and more general concepts to be used in explaining the less general ones.

Finally, there seems to be nothing significantly new concerning two of Gödel's most important features characterizing the objectivity of mathematics: its analytical, but not tautological, nature, and its deep analogy with physics, in that both sciences have to postulate entities from the viewpoint of their fruitfulness. For Gödel, objectivism is clearly a fruitful position for research, and it cannot be adopted just because of its fruitfulness—through some sort of *as if* attitude—for in that case the position is adopted precisely because it has been shown as fruitful (p. 239).

9.6 Philosophy of Logic and Set Theory: Conceptivism

Both in his published and unpublished essays, Gödel spoke rather obscurely about concepts; he seemed to think that concepts are indispensable for philosophy, foundations of mathematics, and logic, and that only through the right concepts the usual paradoxes and other problems of set theory were able to be overcome. Yet not even a minimal theory of concepts and its relations with sets, in the field of a philosophy of logic and mathematics, can be found in those writings. One the novelties of the publication of these records is that here we can finally find some pieces of such a theory. That is why this section will contain more quotations.

The crucial difference between logic and mathematics is that while logic deals with concepts, mathematics deals with sets:

The subject matter of logic is intensions (concepts); that of mathematics is extensions (sets). Predicate logic can be taken either as logic or as mathematics: it is usually taken as logic. The general concepts of logic occur in every subject. A formal science applies to every concept and every object. There are extensional and intensional formal theories... Mathematicians form and use concepts, but they do not investigate generally how concepts are formed, as is to be done in logic. (p. 274)

The subject matter of mathematics is sets:

A set is a unity of which its elements are the constituents. It is a fundamental property of the mind to comprehend multitudes into unities. Sets are multitudes which are also unities... : this is the main fact of mathematics ... Mathematical objects are not so directly given as physical objects. They are something between the ideal world and the empirical world, a limiting case and abstract. Objects are in space or close to space... Sets are quasi-spatial (p. 254).

> Mathematicians are primarily interested in extensions and we have a systematic study of extensions in set theory, which remains a mathematical subject except in its foundations. (p. 274)

On the contrary, logic deals with concepts although it also deals with sets from the viewpoint of its foundations, that is, from the formal side:

> Set is a formal concept If we replace the concept of set by the concept of *concept*, we get logic. The concept of concept is certainly formal and, therefore, a logical concept. But no intuition of *this* concept, in contrast to that of set, has been developed (p. 267). Logic is the theory of the formal. It consists of set theory and the theory of concepts. The distinction between elementary (or predicate) logic, non-elementary logic, and set theory is a subjective distinction... Elementary logic is the logic for finite minds. If you have an infinite mind, you have set theory. (p. 268)

In this connection, we find more about the relation between sets and concepts than what was offered in the ontology (see Sect. 9.3 above). Sets seem to be the extensions of concepts, very like in Freges way:

> It is not in the ideas (of *set* and *concept*) themselves that every set is the extension of a concept. Sets might exist which correspond to no concepts. The proposition "for every set, there is a [defining] concept" requires a proof. But I conjecture that it is true. If so, everything (in logic and mathematics) is a concept: a set, if extensional; and a concept (only) otherwise (p. 274). It is not evident that every set is the extension of some concept. But such a conclusion may be provable once we have a developed theory of concepts and a more complete set theory. While it is an incorrect assumption to take it as a property of the concept of *concept* to say that every concept defines a set, it is not a confusion to say that sets can be defined by concepts or that set is a certain way of speaking about concept. (p. 276)

We can try to go beyond the concept of concept, but to do that we need first a theory of concepts (the one we also need for another important reason: to overcome the paradoxes involved in an incorrect use of them, as it is already suggested in Gödel's publications):

> If you introduce the concept of *concept*, the result is still logic. But going "higher" would be too abstract and no longer logic. The concept of *concept* calls for only the lowest level of abstract intuition... The older search for a satisfactory set theory gives way to a similar search for a satisfactory theory of concepts that will, among other things, resolve the intensional paradoxes. Quines idea of stratification is arbitrary, and Churchs idea along the line of limited ranges of significance is inconsistent in its original form and has not been worked out. (p. 268)

Such a theory of concepts is even suggested along very general lines:

> Even though we do not have a developed theory of concepts, we know enough about concepts to know that we can have also something like a hierarchy of concepts (or also of classes) which resembles the hierarchy of sets and contains it as a segment (p. 278). Just as set theory is formulated in the predicate calculus by adding the membership relation, concept theory can similarly be formulated by adding the relation of application: a concept A applies to something B (which may also be a concept), or B participated in the idea A. Logic studies only what a concept applies to. *Application* is the only primitive concept apart from the familiar concepts of predicate logic with which we define other concepts. (p. 277)

However, a fundamental difference does exist between set theory and a theory of concepts: "the concept of clearly defined concept is not a clearly defined concept. A concept, unlike a set, can apply to itself. Certainly the concept of *concept* is a concept" (p. 278).

To avoid this problem, we could perhaps reach very general, abstract concepts, going higher and higher well beyond the concept of concept itself, which is already rather close to Kantian well-known ideas. There are a few records on this point as well:

> The general concept of *concept* is an *Idea* [in the Kantian sense]. The intensional paradoxes are related to questions about Ideas. Ideas are more fundamental than concepts... Absolute demonstrability and definability are not concepts but inexhaustible [Kantian] Ideas. We can never describe an Idea in words exhaustively or completely clearly. But we also perceive it, more and more clearly... Ideas cannot be used in precise inferences: they lead to the theory of types. It is a kind of defeatism to think that we have this vague idea which is the very basis of our precise idea. We understand the special concept only because we previously had the general idea. We restrict the general idea to individuals to get the concept of the first type. The general idea of concept is just generality. (pp. 268–9)

So, we arrive at the same point we arrived at above, when talking about more and more general concepts from the viewpoint of their "regulative", Kantian use.

9.7 Some Assessment

The book itself has to be highly recommended, in spite of its numerous defects, for a very simple reason: there is no other place where Gödel's philosophical thoughts are available in so frank and open a way. Even a superficial comparison between most of the things Gödel said to Wang and most of the things he wrote in the essays he meant for publication (whether actually published or not), shows that in the first case his extreme caution and pathological fear to controversy disappeared to a very high degree. Also, in these fragments Gödel made several attempts to insert some of his well-known philosophical doctrines into a global framework, both ontological and epistemological.

However, the book is not well-organized, and contains an enormous number of repetitions, while Wangs comments are very uneven in value and perspicuity. It seems that he tried to put into the book all the things he actually had thought about Gödel, even though many of them had already appeared in his former books (Wang 1974, 1987), instead of trying to concentrate just on the new fragments.

Also, the system of numbering, as I said above, is not very informative; even worse, the same system is applied to Gödel's new fragments, to quotations from other materials already published, and even to quotations from other authors. Perhaps it would have been more useful simply to offer the reader the whole list of fragments as they were held by Gödel, together with brief comments by the editor, and then a global attempt to reconstruct the main lines of Gödel's philosophical thought, somehow in the way Everett seems to have attempted first.

As for Gödel's ideas in themselves, I think that many of them are not clearly expressed, so it is very difficult to try to actually insert them into what we already know about his philosophy. For instance, take the concept of concept. He says that the concept of concept is a concept, which is not true with sets, so a theory of concepts similar to set theory cannot be built up. So, it is not clear how a theory of concepts could be built up as a logical, formal foundation of the concept of set, then of set theory, as Gödel seemed to believe, unless we accept explanations of the *obscurum per obscurius* type.

Besides, although Gödel's theory of concepts is based just upon one primitive relation, that of application, this is difficult to relate to Freges theory (from which it could have been inspired). For one thing, Freges concept of concept is not a concept, but an object: this is Freges celebrated paradox of concepts. As we write "the concept of…", we put a concept in an object position, for we use a singular term to name it. Also, for Frege, although the theory of concepts could perhaps be reduced to one fundamental relation, this relation seems to be different from that of application. Frege, in his posthumous writings, wrote about this relation to be that of "subsumption", and this relation is highly problematic, as it may give way to another paradox: the well-known paradox of the relations between terms and the relations actually relating them. That is why Frege invented an ontology of saturated and unsaturated entities (i.e., objects and concepts), which need no further entities to connect them. Yet no similar attempt can be found in the Gödel fragments we now have available.

Finally, Gödel says that concepts could rest on ideas. However, as Gödel himself had to recognize, Kants distinction between ideas and concepts, although useful in trying to define concepts more precisely, "is not clear" (p. 269). So, at the very end, it seems very problematic that a theory of concepts, based on primitive concepts and finally on regulative ideas, could be built up in order to throw light on the most obscure zones of the foundations of logic and set theory when, as it happens, concepts and ideas themselves are very far from being clear entities.

Although this book should be studied by scholars interested in Gödel's ideas, one thing is already clear: we know now why Gödel was so reluctant to make his most deep philosophical ideas public: they are simply rather obscure and rather unconvincing.

References

Kurt Gödel, *Collected Works*, vol. III, Solomon Feferman et al. (eds.), New York and Oxford: Oxford University Press, 1995.

Kurt Gödel, *Unpublished Philosophical Essays*, F. Rodríguez-Consuegra (ed.), Basel/Boston: Birkhäuser, 1995. Original Spanish version: *Ensayos inéditos*, F. Rodríguez-Consuegra (ed.), Barcelona: Grijalbo-Mondadori, 1994.

Hao Wang, *From Mathematics to Philosophy*, New York: Humanities Press, 1974.

Hao Wang, *Reflections on Kurt Gödel*, Boston: MIT Press, 1987.

Departamento de Lógica y Filosofía de la ciencia
Universidad de Valencia

Chapter 10
Propositional Ontology and Logical Atomism

Abstract In the following, I will briefly indicate the role of propositional functions in *Principia,* then point out the way in which they continue to be introduced through propositions, in spite of the doubtful status of these pseudo-entities. I will then try to see if the notion of judgment, which is used by Russell to explain propositions, can meet the requirements which are needed in a serious, coherent ontology. After a survey of the different attempts to build up a convincing notion of proposition and judgment carried out by Russell between 1910 and 1918, I will conclude that the ontology of logical atomism was finally a failure, and so that the mathematical ontology usually associated with Russell's logicism was also a failure. In doing so, I will often show that almost all the problems in which Russell was involved in that period are somehow dependent upon what I will call Bradley's paradox of relations, according to which we cannot consider relations (or other similar "incomplete" notions) as genuine terms, or logical subjects. If we do that, then we should give an account of the way in which these relational terms are in turn related to other terms. Yet this leads to an unavoidable infinite regress.

Keywords B. Russell · Logic · Metaphysics · Philosophy

Mathematics Subject Classification Primary 00A17 · Secondary 00A30

"Logical atomism" is perhaps an appropriate expression far referring to the most important philosophy developed by Bertrand Russell, which is associated with his most important contributions to logic, foundations and philosophy of mathematics. These contributions were mainly presented in *Principia Mathematica* (Whitehead and Russell 1910). I will regard logical atomism as extending, broadly, from 1910 to 1918.

The ontology held by Russell in former periods of his philosophy (mainly in *The Principles of Mathematics* Russell 1903) consists of terms (logical subjects), concepts (including properties and relations) and propositions, these last being genuine entities, possibly to be regarded as a special sort of terms. Here, classes are terms and are used to define numbers and the rest of mathematical entities. The

F. Rodríguez-Consuegra, *Russell, Gödel, Tarski*, Studies in Universal Logic,
https://doi.org/10.1007/978-3-032-08945-8_10

status of propositional functions should, on this scheme, be equivalent to that of concepts, and they, despite their being genuine primitive notions, could easily be introduced in terms of (albeit not reduced to) propositions, given that a propositional function, when a value is assigned to its variable, is transformed into a proposition (as its final value).[1] The problem in logical atomism, beginning in *Principia* and even earlier, is that propositions (in the sense that they are expressed by sentences) are no longer regarded as genuine entities, but merely as sets of entities, so the symbols used to denote them are no longer true symbols, but incomplete symbols. So, propositional functions can hardly be explained through propositions, unless we first clarify the ontological status of propositions.

In *Principia* propositional functions, together with the usual quantifiers, are used to define-eliminate: descriptions; classes, through the axiom of reducibility (which is also named "axiom of classes"); relations, through the corresponding axiom of reducibility (or "axiom of relations"), and also the fundamental notions of identity and membership. From the viewpoint of mathematical ontology, that means that numbers, the basic notion on which mathematics is founded, are defined as classes of classes. And, since classes are nothing but their members and everything that you may want to say about classes can be said through propositional functions, numbers, as well as the rest of the mathematical entities, can legitimately be reduced to propositional functions. Yet, propositional functions are nothing but properties, so numbers are ultimately properties.

Let us first examine how Russell describes the relationship between propositional functions and propositions: "By a 'propositional function' we mean something which contains a variable *x*, and expresses a *proposition* as soon as a value is assigned to *x*" (Whitehead and Russell 1910): vol. I, 38). So, propositional functions cannot even be understood without appealing to the notion of proposition. When you assign a value to the variable in a propositional function ($\phi\,\hat{x}$) you obtain, as the value of the function, a proposition (ϕx). Since propositional functions are genuine entities, and since the values you assign to their variables are entities too [values which belong to the class $\hat{x}\,(\phi x)$], you accordingly have to obtain entities as final values. That is, you can have sentences as values of propositional functions if these are just expressions; you can have propositions as values if propositional functions are treated as entities, but you cannot have propositional functions as entities and sentences as values. Therefore, propositions should be true entities, and according to Russell, they are denoted by propositional functions. Although denotation was interpreted merely as a relation between words and entities (not as a logical relation between concepts, as it was in *Principles),* propositions should be seen as "objects", as Russell writes:

The function itself, $\phi\,\hat{x}$, is the single thing which ambiguously denotes its many values; while ϕx, where x is not specified, is one of the denoted objects, with the ambiguity belonging to the manner of denoting. (Whitehead and Russell 1910: vol. I, 40)

[1] For the details of the whole construction in *Principles,* see my Rodríguez-Consuegra (1991).

Yet the problem starts when Russell writes in *Principia* that propositions are not "single" entities but sets of them, which are in need of judgment to reach the necessary unity, judgment being a "multiple" relation. This theory of judgment was held by Russell in unpublished manuscripts as early as 1906 or 1907 (see my Rodríguez-Consuegra 1989). However, it was not admitted in public until the official proclamation in 1910 (Whitehead and Russell 1910: vol. I, 41–44; Russell 1910b: 155–8[2]), where judgment is openly explained as a multiple relation between the mind and several terms which, taken together, constitute a complex as a consequence of the judgment itself:

> Owing to the plurality of the objects of a single judgment, it follows that what we call a "proposition" (in the sense in which is distinguished from the phrase expressing it) is not a single entity at all. That is to say, the phrase which expresses a proposition is what we call an "incomplete" symbol; it does not have meaning in itself, but requires some supplementation in order to acquire a complete meaning. This fact is somewhat concealed by the circumstance that judgment in itself supplies a sufficient supplement, and that judgment in itself makes no *verbal* addition to the proposition. Thus "the proposition 'Socrates is human'" uses "Socrates is human" in a way which requires a supplement of some kind before it acquires a complete meaning; but when I judge "Socrates is human," the meaning is completed by the act of judging, and we no longer have an incomplete symbol. (Whitehead and Russell 1910: vol. I, 44)

Thus, propositions are not single entities, but rather complexes of entities, which are joined through a special relation. However, propositions (sentences) are now regarded as incomplete symbols, for they need the complement of the judgment itself to reach full meaning, and it is unclear whether this suggests that propositions as complexes are, in fact, nothing at all. In principle, the complex, being a mere creation of the mind, would be something subjective, and this would be a sign that Russell was also trying to avoid one of the forms of Bradley's paradox concerning the statement of any form of correspondence.[3]

The problem lies in explaining how it is possible that the (judging) mind is not a part of the complex which is supposed to be the object of the judgment, while it is, at the same time, to be taken as a part of the multiple relation of judging itself. For in this way the mind can impose a certain order among the terms with which it is acquainted, an order which does not exist on its own. Thus, the status of relations as terms, i.e., as genuine realities, is doubtful, given that, although we accede (through acquaintance) objectively to the constituents of the complex, yet the relation joining them is a mental product (the judging relation). And this, again, turns into the problem about the pre-eminence of relations over terms or vice versa.

On the other hand, it is truly surprising that, for Russell, an incomplete symbol (the proposition before being judged by the mind) can be transformed into a complete one (the proposition once judged by the mind); for an incomplete symbol designates

---

[2] Curiously enough, these pages are part of the new essay that Russell wrote to replace the third section of Russell (1907) (which was here eliminated with no explanation), where he still presented the multiple relation theory only as one more possibility.

[3] I.e., the *relation* of correspondence between two things is in need of another relation which in turn relates it to the two things, and so on.

nothing, while a complete one designates something extra-linguistical. So, the mind seems to have the incredible capacity to transform nothing into something.

Russell's continued his defense in 1907 of the objective status of external relations as the only correct alternative to monism, while he was really espousing, in other manuscripts, the infeasibility of treating propositions as complexes, as well as the subjectivity of the judging relation, which is, in my view, incoherent. As usual, he had to find a compromise between two strong forces pulling in opposite directions. On the one hand, he needed to point out relations as terms[4] to escape from the image he constructed about monism. On the other, he also needed to deny the reality of these relations as terms in one of the most important instances (that of propositions and judgments) in order to avoid Bradley's paradox (now under the form of the impossibility of regarding complexes as terms), while, at the same time, he should avoid any correspondence theory of truth as falling down into another form of the same paradox. Yet some form of such a correspondence was necessary to avoid idealism. But the inconsistencies caused by these two forces appear even more clearly when we consider the details of the multiple relation theory as stated by Russell in *Philosophical Essays* (Russell 1910b) (as in *Principia* he avoids any treatment of such problems).

Let us see how Russell poses the problem of false judgments. In general, the "natural" view concerning judgment says that there is a relation (judging) between our mind and a certain object, the judgment being true or false according to the truth and falsehood of the object of judgment. But Russell had two objections against false objects (Russell 1910b: 175 ff). First, they seem to depend on the full significance that the corresponding sentence acquires through the addition of the expression "I believe that ... ". This is the line leading to judgments as "incomplete symbols". Second, if we admit false objects, we are admitting entities whose existence does not depend upon the existence of judgments, which seems to Russell something incredible, as "it leaves the difference between truth and falsehood quite inexplicable".

Thus, on the one hand, Russell wants to accept some sort of a corresponding entity whose presence or absence can determine the truth or falsehood of the judgment, while, on the other hand, he wants to deny that the absence of this corresponding entity may mean the "existence" of a false object. However, we cannot say that true judgments have objectives while false ones do not, "so long as we hold the view that judgment is actually a relation of the mind to an objective. For... a relation cannot be a relation to nothing" (Russell 1910b: 176–7). The only solution seems dividing the object of judgment into several parts:

> The way out of the difficulty consists in maintaining that, whether we judge truly or whether we judge falsely, there is no one thing that we are judging. When we judge that Charles I died on the scaffold, we have before us, not one object, but several objects, namely, Charles I and dying and the scaffold... We therefore escape the necessity of admitting objective falsehoods, or of admitting that in judging falsely we have nothing before the mind. (Russell 1910b: 177)

Therefore, the "unity of the judgment" cannot be provided by a relation joining the different elements present in it to form a complex, with which the mind is then

[4] Recall that Russell uses the word *term* for *entity* in its most general sense, cf. (Russell 1903: 43).

uniformly related. Rather, that unity is provided by the relation between the mind and each of those elements (the terms and the original relation), "i.e., we must have, not several instances of a relation between two terms, but one instance of a relation between more than two terms". This was the solution Russell discovered to the Bradleyan problem of reconstructing complexes. The later official formulation is this:

> Every judgment is a relation of a mind to several objects, one of which is a relation; the judgment is *true* when the relation which is one of the objects relates the other objects, otherwise it is false... Let us take the judgment 'A loves B'. This consists of a relation of the person judging to A and love and B, i.e., to the two terms A and B and the relation 'love'... The 'corresponding complex' object which is required to make our judgment true consists of A related to B by the relation which was before us in our judgment. (Russell 1910b: 181, 183)

In this way, we can certainly avoid false objects, while some form of correspondence is maintained, although obviously at a rather high price. For, on the one hand, the unity of the judgment seems to depend upon some sort of mental operation which is in charge of ultimately joining the terms and the relation in the judgment; but, on the other, the new theory ignores the ontological advantages of not distinguishing between judgment and perception in the way of Moore,[5] which is involved in Moore's rejection of the correspondence theory of truth. Finally, there is also the problem of the "sense" of the relation between the different terms in the judgment, which has to be related, in some way, to the mind. It may seem that Russell, in discussing these problems, was somehow uncertain of his new theory.

Russell's doubts concern precisely the most difficult point: the relation between the two complexes involved, i.e., the relation between the two relations which give these complexes their unity.[6] Russell writes: "judgment is a relation of the mind to several other terms: when these other terms have *ínter se* a 'corresponding' relation the judgment is true; when not, it is false" (Russell 1910b: 153). A first sign of Russell's doubts is the fact that the relevant correspondence appears with inverted commas. A second sign is the fact that the problem of the "sense" of the involved relations is avoided, while Russell says only that this sense has to be the "appropriate" one:

---

[5] Concerning this last point, Russell says:

> One of the merits of the above theory is that it explains the difference between judgment and perception, and the reason why perception is not liable to error as judgment is... Thus in perception I perceive a single complex object, while in a judgment based upon the perception I have the parts of the complex object separately though simultaneously before me. (Russell 1910b: 181–2)

However, the old Moorean realism made it impossible to make a distinction between fact, or complex object, i.e., perception, and judgment. Thus, any perception involved a judgment (in the sense of a proposition), which made it impossible to maintain any theory of correspondence.

[6] That is, r_1 (mind, A, r_2, B) being the main complex, and where $r_2(A, B)$ is the secondary one, how are r_1 and r_2 related to one another?

> We may distinguish two 'senses' of a relation according as it goes from A to B or from B to A. Then the relation as it enters into the judgment must have a 'sense', and in the corresponding complex it must have the same 'sense'. Thus the judgment that two terms have a certain relation R is a relation of the mind to the two terms and the relation R with the appropriate sense: the 'corresponding' complex consists of the two terms related by the relation R with the same sense. The judgment is true when there is such a complex, and false when there is not. (Russell 1910b: 183–4)

This point is very important for it involves precisely the connection between the two relations, and as one of these relations is mental and the other objective, the danger of idealism seems undeniable. Perhaps such an unsolved problem was the reason why Russell said that his theory preserved the necessary "mixture of dependence upon mind and independence of mind" (Russell 1910b: 158).

Later in the same work, Russell stresses the same point by saying that the new theory succeeds in preserving truth and falsehood as properties of judgments. Thus, they are properties which, in a certain sense, depend upon the existence of minds, while, on the other hand, "the truth or falsehood of a given judgment does not depend upon the person making it or the time when it is made, since the 'corresponding' complex, upon which its truth or falsehood depends, does not contain the person judging as a constituent" (Russell 1910b: 184). Yet I cannot help feeling that in this way Russell comes very close to some sort of contradiction: truth and falsehood of judgments depend upon the mind and then again do not depend upon it. This might be one of the consequences of having forgotten Moore's radical way of avoiding the problem, which consisted in rejecting any correspondence theory by identifying judgment with perception.

In *Problems of Philosophy* (Russell 1912b) Russell already explicitly considered the problem of the sense of the relations involved but provided no solution to it. The starting point is already misleading, for in pointing out several examples of multiple relations, he refers only to multiple relations among people, while the example which is actually analyzed later is one in which one of the terms is a relation: "Othello believes that Desdemona loves Cassio". Russell then admits two different senses, according to the two relations involved, but he continues to maintain that the judging relation provides its sense to the secondary relation: "the relation 'loving', as it occurs in the act of believing, is one of the objects – it is a brick in the structure, not the cement. The cement is the relation 'believing" ' (Russell 1912b: 128). However, this can only mean that the terms are put in a certain order precisely "by the 'sense' of the believing" "(*ibid.*), with which, if they have no order by themselves, it is difficult to see how it is possible to speak of the 'correspondence' between the two complexes.[7]

---

[7] There is another indication of the extreme importance of the problem of sense in Russell's correspondence with Broad. See for instance this letter to Russell of June 25, 1912:

> Suppose we have an n-adic relation. Then the number of possible [?] complexes of the same terms will be |n. In a judgement about such a complex we shall have $n + 2$ terms, i.e. [?] the n terms of the original relation, the relation itself, and the judging mind. So that believe will involve a $n + 2$-adic relation and therefore |n + 2 psychical states differing only in the sense of their relating relation will be possible. If for a true belief each complex of the object terms must be correlated with one and only one of the psychical complexes then ought to be for

Moreover, the problem concerning idealism remains unaltered, as does the problem concerning Bradley's paradox, in both forms of the endless regress: clarifying the relation between relation and terms, and the complex indicating the correspondence between the two previous complexes. If the judging relation imposes a sense on the secondary complex, then this sense is a product of the mind, given that the mind is one of the constituents of the judging relation. On the other hand, if the mind has acquaintance with each member of the complex,[8] then it has to be acquainted with the *relation* as well, and it seems that this should include the involved sense. If not, not only does the judging relation impose its own sense on the secondary complex (with which we might produce "nonsense", as Wittgenstein later objected), but also we must explain how it is possible to *connect* this second relation to the

---

> any given complex of objects $\underline{|n+2} - 1$ possible false beliefs as to the sense. But actually there are only $\underline{|n} - 1$. Would you say that $\underline{|n+2} - \underline{|n}$ of the possible psychical complexes are not beliefs, or that as a matter of fact they never exist? If so on what principle do you decide what are beliefs or which exist? And, apart from this, can we tell [?] at all which sense of the object complex is correlated with which one of the belief complex? Finally the relation of belief in different beliefs has different polyadicity according to that of the relation of the object complex. Is there [?] any other example of the same relation having different polyadicity? I think you hold that relations do not have instances; what then is there in common between all the differently polyadic relations that relate the complexes which are all called beliefs?

And Russell's response of Jan 31, 1912:

> It is not the case that $\underline{|n+2}$ psychical states differing only in the *sense* of the relation of judging are possible. A mind must occupy one fixed position in the complex. E.g., "B. judges that the sun is shining" is possible, but not "the sun judges that B. is shining". If several of the other constituents of the judgments are minds, of course the case is altered [?]. But this is only possible when God is the Judge, since we have no acquaintance with other minds. And then polytheism is required really. Generally, your principle that $\underline{|n}$ complexes can be made of *n* terms is wrong; many senses will not yield a complex if the terms are of different sorts.
>
> The question whether we can tell which sense of the object-complex is correlated with which of the belief-complex is more difficult. It is plain to me that we can, but I hardly know how. I think it must come by way of acquaintance with a perceived complex. You perceive "A-to-right-of-B" and you judge that *A* is to right of *B*; this gives a correlation of the two senses. Thence it could be extended to cases when the complex is not perceived. But I am not sure that this is satisfactory.
>
> The question of the different polyadicity of different relations of belief is difficult too. I have been always inclined to suppose these [?] were different relations of belief, 3-term, 4-term, etc. But then, as you say, one wants to know what they all have in common, and to that I don't know what to answer. Perhaps only an associated feeling. The question is serious and I should be glad to know the answer.

Both letters are extant in the Russell Archives, McMaster University, Hamilton, Canada.

[8] In Russell (1912b) Russell openly admits this point: "when we are judging, we have *a* relation to each of the constituents of our judgment separately" (Russell 1912b: 153), but he avoids the term "acquaintance" and replaces it by "being conscious of", perhaps to make the problem seem less important.

related terms, with the subsequent danger of arriving at Bradley's regress in the first form.[9]

The second form is unavoidably implied by the supposed 'correspondence' between the two complexes. We must not forget, first, that for Russell, at this stage, there are no propositions per se, for they (the corresponding sentences) are mere incomplete symbols which exist only in judgment (or beliefs, which are synonymous for Russell in this context). Thus, there is no complex $A\ r\ B$ on its own – there is only "I belief that $A\ r\ B$." At this point the possibility of the correspondence is quite obscure, for as it is the judging relation that imposes its sense, thus actually producing a complex (which does not exist independently), it is hard to accept that there must be a correspondence between the judging relation, which is a product of the mind, and the resulting secondary complex, which is also a product of the mind, if the theory of incomplete symbols is to be accepted. The endless regress is present here, just as it is in any other form of correspondence: to explain the relation of correspondence between the first complex, which is a belief (the judging relation), and the second complex (the secondary relation), we need another belief stating the fact of the correspondence, as Russell wrote many times when he was attacking the correspondence theory at former stages of his philosophy.

However, here the second kind of endless regress can be presented even under another more subtle form, which was already suggested by Stout in a further criticism (Stout 1915), in spite of the fact that Stout did not see any problem of the form of a Bradleyan endless regress. The fact is, just as there must be a correspondence between the two complexes, we also have to know whether or not the judgment-complex is itself apprehended, for we must *compare* it with the secondary complex. But "this would imply that whenever we believe we must at the same time be aware of the state or process of believing, and of the mind as a constituent of it" (Stout 1915: 343). In other words, if we have "I believe that $Ar_1\ B$", and we need to compare the two complexes "r_2(Mind, A, r_1, B)" and "$A\ r_1\ B$" to know about the possible correspondence and then about the truth of the judging relation, then we also need "I believe that { r_2 (Mind, A, r_1, B)} r_3 {$A\ r_1$ B}", and so on.

The climax of all these problems can be found in the articles in which Russell considered the status of relations and predicates, in an attempt to set up a whole ontology in which a full classification of universals and particulars was provided, once the multiple relation theory had been publicly admitted. In those articles we can see how the inconsistencies between the two forces we have described above come close to being full-fledged contradictions. In the following, I shall point out some of these inconsistencies in the four main papers published between 1911 and 1912.

---

[9] Anyway, it is also unclear how we are to understand Russell when he says that the sense is ultimately provided by the judging relation, for the real problem in jealousy does not, I am afraid, lie in whether Desdemona loves Cassio or Cassio loves Desdemona, but rather in whether or not Desdemona loves Cassio. And this indicates that the mind can not only provide a sense, but also produce a relation which does not exist *relating Desdemona and Cassio.* (Although perhaps this is not quite the same as saying that the mind creates the relation at all, for Russell believed that we have acquaintance with universals.)

In Russell (1911a) we first find the argument that we have "awareness" of universals, and especially of relations, although the "proof" is not very convincing: "yellow differs from blue". But treating relations as things that are immediately known to us supposes that their recognition has an objective ontological status which, although derived ultimately from Russell's need for regarding them as genuine terms (then as ultimate realities), is hardly compatible with *Principia* where relations, like classes, are regarded as mere fictions, and their names rendered as incomplete symbols accordingly. It can be said that the relations which were considered there were only relations in extension, but to say that is not to say much, for classes had the same extensional status, and Russell never tried to admit them as objective realities "in intension" in his ontology, while they had to be rejected in logic, which is extensional (except, of course, through propositional functions, which, being ultimately properties, are intensional entities).

This would be a constant ambiguity in the rest of Russell's development, and I think it ultimately proceeds from the tension between the two forces we have been considering since the beginning of this paper. The only argument we are given is that we not only directly know complexes that contain relations as constituents, but also relations as logical subjects, which is a consequence of the principle of acquaintance, according to which any proposition we can understand must be composed of constituents with which we are acquainted (Russell 1911a: 117). However, the principle presupposes that there is already a solution to the problem of relations as terms, which, as we have seen, can hardly be maintained.

Another sign of Russell's problems with relations regarded as "forms" involves the difficult status of propositional functions themselves. In *Principia,* as we have seen, they were to be regarded as primitive ideas to which all the rest of our "formal" concepts (i.e., classes, relations, descriptions, etc.) could be reduced. Here he adds that propositional functions are "complexes" which play the role of "true subjects" or "ultimate subjects" (Russell 1911a: 126, 128), despite the fact that this would suppose (i) that we would have acquaintance with them, and (ii) that they would be true constituents of complexes, as they appear in judging multiple relations. Of course, Russell needed to regard propositional functions as logical subjects in *Principia,* but he provided no theory to explain these two consequences. This, together with the problems pointed out above, might have contributed to the changes in the multiple relation theory of judgment in *Theory of Knowledge*[10] where "forms" are openly admitted as being parts of the multiple relation, as we shall see below.

The essay (Russell 1911b) tries to build up a whole philosophical view, epistemological and ontological at the same time, but the problems of relations as terms remain unsolved. Russell starts by writing:

> It [my philosophy] is analytic for it maintains that the existence of what is complex depends upon the existence of what is simple, and not vice versa, and that a constituent of one complex is absolutely identical, as constituent, to what it is in itself when its relations are not considered. (Russell 1911b: 53)

[10] Written in 1913, but not published until 1984. See below.

However, Russell seems to forget that relations are also constituents of complexes, so that some explanation of their status as simples must be provided, especially because it is unclear how it is possible that relations are exactly the same as the rest of the constituents of a complex, i.e., how it is possible to regard relations both as relating relations and as mere constituents that are not related at all. The problem again has to do with the twofold nature of relations: the paradoxical consequences we have been seeing force us to recognize them simultaneously as terms and concepts.

That is why when Russell classifies the different kinds of "beings" in the world, he is silent about the place where relations are to be found: "I say then that there are simple beings in the universe, and that these beings have relations through which they compose complex beings" (Russell 1911b: 56). He admits that every complex has two kinds of constituents: terms and relations (or predicates). So, he seems to have forgotten that, according to his own view, every complex must also contain relations *as terms,* so the distinction is difficult to accept. Finally, since Russell admits predication as a true universal, he should provide some explanation of the ontological status of predication, and even of the consequences concerning the old theory of judgment, according to which there is no ontological difference at all between subject and predicate. But nothing of the kind is to be found in this paper.

The paper (Russell 1911c) tries to provide some responses to these problems. We have to remember that Russell needed, at the same time, to consider predicates as relations (or as involving relations, which give their true essence) and also as predicates in themselves. He needed the former, because in this way some form of the old Moorean theory of judgment could still be maintained (and so the danger of the predicative general form avoided); but he needed the latter too, because propositional functions, the most important entities in *Principia,* are nothing but the general form of properties (and therefore they seem to be able to reduce all propositions to the predicative form).

In this paper Russell introduces the difference between monadic and dyadic concepts, doubtless in an attempt to face the difficulty, but is then forced to speak of predicates as having the two different natures: "It is of course the case that, whenever a subject has a predicate, there is a dyadic relation of subject and predicate, but it does not follow that there is not also a proposition in which the predicate is merely predicated" (Russell 1911c: 159). That implies further that the analogy with relations is complete: we have predicates as relations ("predicating" predicates) and predicates as terms (predicates "in themselves"), which leads exactly to the same unsolved problems as with relations (for we also have "relating" relations along with relations "in themselves"). The first form avoids Bradley's paradox but cannot provide terms, and the second form provides true terms but cannot explain complexes, unless we introduce further relating constituents and accept Bradley's paradox. The following passage is a further effort to make sense of the distinction:

> Whenever *a* has the relation *R* to *b,* there is a triadic relation of *a* and *R* and *b,* but in this relation *R* occurs as a term of the relation, not as the relating relation of the proposition. Similarly, if there are monadic concepts, the proposition in which they are said to have the relation of predication to their subjects will not be identical with the propositions in which they are actually predicated. (Russell 1911c: 159)

But the new maneuver is only the old strategy already used to avoid Bradley's paradox, for here we have two additional relations which do not appear in the notation: one is the triadic relation which actually relates the rest of the constituents; the other is the relation R no longer as a relation, but as a term. Thus, to explain a dyadic relation we need a triadic one where the former relation acquires a different status; so presumably to explain the triadic relation we are going to need a quadruple one, and so on. Since the same goes for predicates, we get the same problem Russell encountered in former stages: there is no point in asking whether or not predicates can be transformed into relations, for predicates are already relations.

The paper (Russell 1912a) can be regarded as the final stage of the series. Its main goal was to defend the ultimate dualism between universals and particulars, which Russell needed in order to separate concepts (predicates and relations) from terms. But he also needed concepts to be terms as well, which made any consistent solution impossible.[11] Thus, the classification itself is vitiated by the old problem of the twofold role. That is why when Russell says that particulars enter into complexes only "as the subjects of predicates or the terms of relations", and universals "as predicates or relations" (Russell 1912a: 124), he seems to be oblivious of the fact that, according to his own position in other contexts, predicates can be subjects and relations can be terms.

Here Russell openly admits that predication is a relation "involving a fundamental logical difference between its two terms" (Russell 1912a: 123), but this is not much, for at the same time he needs to admit the twofold role of predicates, which makes the supposed logical difference impossible. As Russell himself confesses, he needs to maintain a specific relation of predication to be able to make an ultimate distinction between particulars (those which cannot be predicates or relations) and universals (those which are only predicates or relations). But as we have seen the distinction cannot be ultimately maintained, for it amounts to nearly the same thing as saying, in a language close to Bradley's, that universals are already impregnated with particularity. At this stage Russell was no longer able to accept such contradictory metaphors.

I do not know how far these efforts can be located in relation to Bradley's theory that universals are also particulars and vice versa, but, at any rate, it seems that the corresponding idealism is somehow involved in Russell's view. We will also discover that the monism (and the holism) implicit in this idealism will appear more and more openly in later stages.

Anyway, the polemic which Russell maintained with Bradley directly, both in the publications and in the unpublished correspondence, can be usefully considered in this context. Bradley started the (published) polemic in [1], by accusing Russell of maintaining an inconsistent pluralism, for he admitted "unities which are complex and which cannot be analysed into terms and relations" (Bradley 1910: 176). Russell's response in Russell (1910a) tried to avoid the ultimate inconsistency by saying that

---

[11] In *Principles* Russell juggled with the problem by introducing a further division of terms, into things and concepts. But then the problem remains when we have to give an account of the fact that relations—concepts in principle—can also be regarded as things.

he did not maintain that unities are incapable of analysis, but only that the mere enumeration of the constituents cannot reconstruct the unity: "A complex differs from the mere aggregate of its constituents, since it is one, not many, and the relation which is one of its constituents enters into it as an actually relating relation, and not merely as one member of an aggregate" (Russell 1910a: 344). But in saying this he provides no true reply to the objection, as Bradley lucidly pointed out in Bradley (1911):

> Is there anything, I ask, in a unity beside its "constituents", i.e., the terms and the relation, and, if there is anything more, in what does this "more" consist? Mr. Russell tells us that we have got merely an enumeration or merely an aggregate. Even with merely so much I should still have to ask how even so much is possible. But, since we seem to have something beyond either, the puzzle grows worse.

The personal correspondence[12] shows Russell to be much more appreciative of Bradley's philosophy. In a letter from 1907, Bradley had already pointed out that wholes can by no means be reconstructed, for they are non-relational in the last analysis: "But if you break this entity up, and set down any part as independent—then, starting with this part, there is no getting beyond it except arbitrarily… You will say that you replace this by externa! relations. But it is denied that these serve" (Oct. 21, 1907). Russell admitted the difficulties, but only by claiming that relatedness does not imply complexity, with no further explanation of how we can give an account of relations as *related* to their terms (Oct. 29, 1907).

The final stage of the correspondence is captured very nicely in this passage by Russell, which contains the failed attempt to escape the old paradox: "I do not consider pluralism incompatible with the existence of complex entities. I consider that in every case where two simples have a relation, there is a complex entity consisting of the two simples so related" (April 9, 1910). But this, again, misses the point, for the expression "complex entity" involves different unexplained senses. We only have to replace "simples" with "terms" to realize that Russell is forgetting that the relation is also a term on his own view. This in turn requires him to explain the difference between the complex as formed by the three terms, and the complex as being the "complex entity", supposedly formed by the terms as "simples" and the "relating relation". This is why Russell finally added a third entity, the "form", and faced the same paradox at a higher level.

Russell 's final attitude implied the recognition that even considering the recent publications from 1910–11, the problem of unities remained unsolved: "I have nothing short to say, the subject is difficult (…), and I do not pretend to have solved all its problems" (March 2, 1911). Fortunately, there is still another passage showing the strong link between Bradley's criticisms, Russell's unsolved difficulties, the new theory of judgment, and Wittgenstein's objections (Jan. 30, 1914):

---

[12] In the Russell Archives, McMaster University, Hamilton, Canada. A more detailed study of this correspondence will appear in my *Relational Ontology and Analytic Philosophy,* now near completion. [This book was finally published, with the subtitle *Bertrand Russell and Bradley's Ghost*, as part of the digitization of my Academic Archives series, vol. 4, in 2019, and is available here: https://www.amazon.com/dp/1709974710.]

> I fully recognise the vital importance of the questions you raise, particularly as regards "unities"; I recognise that it is my duty to answer if I can, and if I cannot, to look for an answer as long as I live... Chiefly through the work of an Austrian pupil of mine, I seem now to see answers about unities; but the subject is so difficult and fundamental that I still hesitate.

To my knowledge, this is the only place where Russell admitted these important links, as well as the fact that he regarded his former views as a failure, precisely from the viewpoint of Bradley's objections, and not only as regards Wittgenstein's. The next stage, which might perhaps be entitled: "Enter Wittgenstein: the multiple relation theory staggers on", can, I think, be regarded, to some extent, as the time at which Russell's philosophy had to pay the price for his continuous delay in facing the "fundamental principles" (as Bradley used to say). This price turned out to be quite high, for it involved the abandonment of a major project in the philosophy of logic and epistemology, and his subsequent devotion to particular problems, without any hope of finding an acceptable global philosophy. However, the difficulties which made the project impossible were already present in former stages, as already pointed out by Bradley, despite the fact that Wittgenstein's criticisms were given the credit for Russell's disappointment with his own philosophy.[13]

The manuscript (Russell 1912c) is already an attempt to characterize the notion of form. But the starting point returns to former views: "The *form* of a complex is what it has in common with a complex obtained by replacing each constituent of the complex by something different".[14] Thus, we have to avoid Bradley's paradox, for if we make the form a constituent, "it would have to be somehow related to the other constituents, and the way in which it was related would really be the form; hence an endless regress" (Russell 1912c: 2). Therefore, though no final definition of the notion is provided, Bradley's paradox is respected, and for the same reason any possible violation of the theory of types was apparently avoided, for any attempt to consider forms on the same ontological level as constituents would be such a violation, for it would regard "formal" concepts as individuals or substances.

That is why it is so strange that in the unfinished book of 1913, *Theory of Knowledge* (Russell 1913), Russell's main recourse was the admission of forms as constituents of complexes with no explanation of the type-theoretical problems involved. Of course, this was also an attempt to avoid the criticisms of Bradley and Wittgenstein, which forced him to introduce some changes to avoid the rather idealist consequences of regarding complexes, and then propositions, as a mere creation of our minds. But these criticisms forced Russell to face a very unpleasant dilemma: he could either maintain the old theory of types by abandoning the multiple relation theory of judgment and by renouncing the attempt to characterize complexes, and then logic, or abandon some philosophical consequences of the theory of types by complementing the multiple theory with an explicit device making the admission of forms as some sort of constituent possible. Russell's solution in 1913 was clearly

[13] Griffin's (1986) also contains a good survey of those criticisms.

[14] The examples which Russell mentions are propositional functions, dyadic and multiple relations, and the two standard forms of quantification.

a compromise between the two horns of the dilemma, although the final solution seems to involve the second alternative, despite Bradley's prohibition.

Russell now rejected the view that the form can be a "mere" constituent of complexes with the usual Bradleyan argument. Hence, in "Socrates is human," "is" represents the form, and thus cannot be a constituent: "for if it were, there would have to be a new way in which it and the two other constituents are put together, and if we take this way as again a constituent, we find ourselves embarked on an endless regress" (Russell 1913: 16). But when he tries to explain the nature of form, he rejects regarding it not only as an equivalence relation (i.e., "to be the same form"), but also as a mere primitive idea; for it would lead us to the usual paradox, when trying to relate this idea with the others in the system. The solution, already found in Wittgenstein's framework, was to regard it as an indefinable object corresponding to certain general expressions.

Thus, the form of subject-predicate complexes will be "something has some predicate," and the form of dyadic complexes will be "something has some relation to something." Russell tries to avoid the obvious attack of circularity by writing: "in spite of the difficulties of language, it seems not paradoxical to say that, in order to understand a proposition which states that *x* has the relation *R* to *y*, we must understand what is meant by 'something having some relation to something" ' (Russell 1913: 114). However, the compromise supposes only one change in the general scheme of the already published multiple relation theory of judgment: incorporating a symbol of the form (γ) into the general complex constituting the judgment: "$U(S, x, R, y, \gamma)$". Thus, it is difficult to deny that forms are regarded as constituents. However, this is not only an attempt to leap over Bradley's paradox, but also to give an objective status to the form as something with which we are acquainted, in the same way as a relation was both a relating relation and a term. The only change is now to regard γ as the general form of dual complexes, and to show it to absorb the "relating" part of *R*, though, of course, we still need to explain the status of γ in the complex.

I cannot enter here into Wittgenstein's criticisms in any detail[15] which forced Russell to leave the book unpublished, but for Wittgenstein the main defect of Russell's revised theory was that all the difficulties involved had their common root in the attempt to regard the form as a new constituent. Wittgenstein expressed this in a language shrouded in mystical connotations, but this cannot cover the fact that he clearly saw the impossibility of Russell's attempt to fix the nature of a form without simultaneously using another form. Thus, Bradley's paradox is also present: we cannot meaningfully *speak* about the relation between the judging complex and the secondary complex.

If we look at his publications at this stage, Russell, rather surprisingly, did not abandon the multiple relation theory of judgment, nor renounced the need to regard forms as constituents. In Russell (1914) Russell starts by declaring forms to be the object of "philosophical logic" (Russell 1914: 52), although in resorting to the "replacement" device, he says only that "form is not another constituent, but is the way the constituents are put together". Thus, without mentioning his failure to

[15] I did it in Rodríguez-Consuegra (2002).

construct an acceptable epistemology of logic, he even adds the explicit claim that we have knowledge of forms, allowing us to understand sentences: "Thus some kind of knowledge of logical forms, though with most people it is not explicit, is involved in all understanding of discourse" (Russell 1914: 53). We have thus the two traditional arguments, but nothing about the status of the form in the judging complex, while the multiple theory is apparently maintained, for Russell denies objective negative facts except as false beliefs: "It is therefore necessary, in analysing a belief, to look for some other logical form than a two-term relation" (Russell 1914: 66). In the face of that statement, I am unable to see any weight in the usual claim that Russell "abandoned" the multiple relation theory in 1913 because of Wittgenstein's criticisms.[16] On the whole, then, I must conclude that Russell made no progress in trying to solve the real problems underlying all of his rather edifying talk of relations.

The last paper (Russell 1956) was Russell's final attempt to maintain a consistent theory before officially abandoning the multiple relation theory of judgment. Yet, we again find exactly the same unsolved problems. As regards forms, the same idea of constituting an inventory is introduced by merely changing "forms of propositions" to "forms of facts" (Russell 1956: 216), which surfaces again as a "realistic bias" despite the need for providing an account of false "facts" in terms of some kind of multiple relation theory. This multiple theory seems *to still be maintained here,* for it is said that belief is not a dual relation: "Therefore the belief does not really contain a proposition as a constituent but only contains the constituents of the proposition as constituents" (Russell 1956: 224).[17] Bradley's paradox is presented a few pages later, when Russell admits that we cannot put the verb on the same level as its terms

[16] Russell seems to have maintained some sort of multiple relation theory in his lectures in America. The following are some notes from V. Lenzen's "Notes on Russell lectures" (now extant in the Russell Archives), which were taken in March 1914, that is to say, after Russell had left *Theory of Knowledge* unfinished and, supposedly, abandoned the theory of judgment it contained. In Lenzen's notes we can read, for example, about propositional attitudes as involving not a dual but a multiple relation:

> Judgment: all objects must be things with which you are acquainted... Acquaintance with universal—logical form of occurrence—not same as acq. with particulars. Possibility of error in any cognitive occurrence shows that occurrence is not dual relation... I believe Jones hates Smith—single fact&—contains 2 verbs. Constitutes oddity of propositional thought... Logical form of occurrence is different from that of presentation.

Lenzen's term paper dealt precisely with Russell's theory of judgment, and it contained two main criticisms: judgment cannot be a relation, for (i) truth *is* a relation; that is why we say there are true judgments; (ii) relations are universals, while judgment is a process in time. Russell's reply, in the form of notes added to the paper, reads: "*Judgment* is a relation, *a judgment* is not a relation. Thus *man* is a universal, but *a man* is not. Your argument (...) on this point sins against philosophical grammar". Also: "'A judgment' will be a positive fact in which the principal relation is *judging*; but a judgment is not itself a relation. What is related to an objective is judgment, not a judgment". Thus, Russell does not confess his strong doubts concerning the multiple relation theory after Wittgenstein's Bradleyan criticisms, but he seems to continue to maintain the theory rather *explicitly.*

[17] For, as stated before, "Every fact that occurs in the world must be composed entirely of constituents that there are, and not of constituents that there are not" (Russell 1956: 220).

because it is an instance of form, which can by no means be a further constituent: "the form of the dual relation… is not a constituent of the proposition. If it were you would have to have that constituent related to the other constituents" (Russell 1956: 239). The problem is, then, the same: Bradley's paradox makes the form as a constituent impossible, but precisely this is required by the multiple relation theory.

The persistence of these ideas can even be seen once again in Russell (1919), written in 1918 (Russell 1919: 198 ff). Here Russell needed some definition of logic[18] for the book was explicitly devoted to the logicist foundations of mathematics. He resorts to the usual "solution": "we may accept, as a first approximation, the view that *forms* are what enter into logical propositions as their constituents," with which "logic is concerned only with *forms,* and is concerned with them only in the way of stating that they are always or sometimes true." Thus, he says not only that forms are constituents, but also that we can have second-order propositions in which we refer to these forms and the rest of the constituents. He is obviously forgetting the two forms of Bradley's paradox. However, a few lines earlier he wrote: "the form is not itself a new constituent; if it were, we should need a new form to embrace both it and the other constituents."

If I have to propose some conclusion here, I can only add that the chaotic philosophy usually called "logical atomism", can hardly be regarded as "logical" or as "atomism". It is not logical, for it was mainly the compromise between two incompatible, purely philosophical (ontological) traditions: that of Moore, and that of Bradley. Neither is it atomistic, for it was mainly devoted to including forms in the ultimate inventory of the world, which was impossible as it was attempted by making forms to have a twofold nature, that of terms, and that of non-terms. Furthermore, this philosophy was hardly realistic, for the unavoidable idealistic implications of the multiple relation theory of judgment, together with the monism implicit in the logical constructions, made it impossible to continue to maintain that complexes are genuine objective realities. They are, at best, the result of *our* application of forms to the terms we perceive in the world. But since these terms, even when they are the simplest possible, can be further constructed out of relational structures, which are ultimately nothing but "incomplete symbols", the old realistic and atomistic world is replaced by an inescapably idealistic and holistic one, where terms can no longer be opposed to relations, where relations can no longer be regarded as true terms,

[18] In a letter of 1918 to Frank Russell Russell says that the most important things still to be reached at that stage were: (i) a theory of judgment; (ii) a definition of logic. The new orientation, as to the *solutions,* is now openly psychological, but the *problem* is still the same: the unity of complexes. I think it is worth quoting the letter:

> "Facts, Judgments, and Propositions" opens out — it was for its sake that I wanted to study behaviourism, because the first problem is to have a tenable theory of judgment. I see my way to a really big piece of work, and incidentally to a definition of "logic", hitherto lacking. All the psychology that I have been reading and meaning to read was for the sake of logic; but I have reached a point in logic where I need theories of (a) judgment (b) symbolism, both of which are psychological problems. (Russell 1919–1926: 249)

and where the seed of the later abandonment of the distinction between subject and object was already present.

Acknowledgements I am grateful to an anonymous referee for useful suggestions. [My thanks were also due to the organizers of the great 2001 Russell Munich Conference for inviting me, where I exposed this paper. Also, a few of these ideas were exposed in the IHPST, Université de Paris I Panthéon-Sorbonne, in April 1998.]

References

Bradley, Francis H.: 1910. On appearance, error and contradiction. *Mind* 19: 154-185.

Bradley, Francis H.: 1911. Reply to Mr. Russell's explanations. *Mind,* 20, 74-76.

Griffin, Nicholas: 1986. Wittgenstein's criticisms of Russell's theory of judgment. *Russell* 5: 132-145.

Rodríguez-Consuegra, Francisco: 1989. Russell's theory of types, 1901–1910: its complex origins in the unpublished manuscripts. *History and Philosophy of Logic* 10: 131- 164.

Rodríguez-Consuegra, Francisco: 1991. *The Mathematical Philosophy of Bertrand Russell.* Basel: Birkhäuser. reprinted in 1993.

Rodríguez-Consuegra, Francisco: 2002. Wittgenstein y la teoría russelliana de la proposición. In my *Estudios de filosofía del lenguaje,* Granada: Comares, 179–214. Partial English translation of this paper, with some changes and new material, as "Wittgenstein and Russell on propositions and forms". In J. Padilla Gálvez (ed.), *Wittgenstein from a New Point of View.* Wittgenstein-Studien, vol. 6, Frankfurt a. M.: Peter Lang, 2003, 79–110.

Russell, Bertrand: 1903. *The Principles of Mathematics,* Cambridge: The University Press. London: Allen & Unwin. Second edition in 1937. Paperback edition London: Routledge, 1992.

Russell, Bertrand: 1907. On the nature of truth. *Proc. Arist. Soc.* 7: 28-49.

Russell, Bertrand: 1910. Some explanations in reply to Mr. Bradley. *Mind* 19: 373–378

Russell, Bertrand: 1910. *Philosophical Essays.* London: Longmans Green.

Russell, Bertrand: 1911. Knowledge by acquaintance and knowledge by description. *Proc. Arist. Soc.* 7: 108–128.

Russell, Bertrand: 1911. Le réalisme analytique. *Bull. Soc. Franç. Phil.* 11: 53–82.

Russell, Bertrand: 1911. The basis of realism. *Jrn. Phil.* 8: 158–161

Russell, Bertrand: 1912. On the relation of universals and particulars. *Proc. Arist. Soc.* 12: 1–24.

Russell, Bertrand: 1912. *Problems of Philosophy.* London: Williams and Norgate.

Russell, Bertrand: 1912. What is logic? In: *The Collected Papers of Bertrand Russell,* vol. 6, London: Routledge, 1992.

Russell, Bertrand: 1913. *Theory of' Knowledge:* the 1913 manuscript. In: *The Collected Papers of Bertrand Russell,* vol. 7, London: Allen & Unwin, 1984.

Russell, Bertrand: 1914. *Our Knowledge of the External World.* London: Allen & Unwin.

Russell, Bertrand: 1918–1919. The philosophy of logical atomism. *The Monist* 28–29. Repr. in *Logic and Knowledge.* Edited by R. Ch. Marsh, London: Allen & Unwin, 1956.

Russell, Bertrand: 1919. *Introduction to Mathematical Philosophy.* London: Allen & Unwin, 1 98ff.

Russell, Bertrand: 1919–1926. The collected Papers of Bertrand Russel, vol 9.

Stout, Georg F.: 1915. Mr. Russell 's theory of judgement. *Proc. Arist. Soc.* 15: 332-352.

Whitehead, Alfred N., and Bertrand Russell: 1910, 1912, and 1913. *Principia Mathematica,* 3 vols. Cambridge: Cambridge University Press.

Departamento de Lógica y Filosofía de la ciencia
Universidad de Valencia

Chapter 11
Is Tarski's Theorem Really Tarski's?

Abstract This is an attempt to clarify the different senses of the expression "Tarski's theorem", according to different presentations. In the most usual one, it's defended that the priority should be attributed to Gödel, rather than to Tarski. The views of Tarski and Gödel are also presented and compared. Some unpublished material is also used.

Keywords K. Gödel · Logic · Mathematics · Philosophy

Mathematics Subject Classification Primary 03–02, Secondary 03Exx.

11.1 The Received View

What is usually known as "Tarski's Theorem" says that arithmetical truth is not definable in arithmetic, or sometimes, more generally, that a truth predicate for the object-language is not definable within the object-language. The expression is used this way in many different contexts, including manuals and historical literature. Also, it is sometimes believed that Tarski's Theorem is one of the theorems which were explicitly stated and proved by Tarski in his celebrated monograph on truth (Polish 1933, German 1935, English 1956a). Besides, although it is usually recognized that both Tarski and Gödel independently discovered the method of arithmetizing the metalanguage, no especial link is usually pointed out as existing between Tarski's Theorem and Gödel's incompleteness theorems from the point of view of the possible dependence between them—historical, logical or otherwise (see for instance Parsons 1974, pp. 75 ff, and Kleene 1975, pp. 52–3). Sometimes it is said (for instance in Wang 1974, pp. 168–9) that the fact that arithmetic truth is not definable in arithmetic was first emphasized in publications by Tarski (1933a, b) and Gödel (1934), but again without establishing any relationship between them of the kinds mentioned (see however below on other writings by Wang).

F. Rodríguez-Consuegra, *Russell, Gödel, Tarski*, Studies in Universal Logic,
https://doi.org/10.1007/978-3-032-08945-8_11

We shall see in the following the precise sense in which all this can be understood, but let me first determine the sense according to which Tarski's Theorem can be found in Tarski's writings. The so called theorem could perhaps be confused with Theorem I in Tarski 1935, §5 (1956a, p. 247), which says: (α) if a truth predicate "denoting a class of expressions, is defined in the metatheory, it will be possible to derive from it the negation of one of the sentences" described in the convention T; (β) if the class of all provable sentences of the metatheory is consistent, "it is impossible to construct an adequate definition of truth in the sense of convention T on the basis of the metatheory". However, Theorem I refers to languages of infinite order, and what it states is rather the impossibility to define truth for an object-language of infinite order even in its metalanguage, although the claim that the truth predicate for the object-language cannot be defined in the object-language is contained in its proof (see Coffa 1987, p. 563, where, in addition, the underlying philosophical reasons are explained). In §7, the Postscript added in the 1935 German edition, Tarski, by renouncing to the earlier "theory of semantical categories", extended the former procedures to languages of any order. Thus what remains is simply that truth is not definable in the object-language considered, but in a stronger metalanguage. This is then applied to the language of arithmetic: "It is impossible to give an adequate definition of truth for a language in which the arithmetic of the natural numbers can be constructed, if the order of the metalanguage in which the investigations are carried out does not exceed the order of the language investigated" (1956a, p. 272).

11.2 More Recent Interpretations

The present situation may perhaps be posed by quoting the two more recent "official" accounts of Gödel's and Tarski's achievements related to truth. In the first case we read that §7 of Gödel (1934), in discussing the relation of his incompleteness results to the paradoxes, contains "the indefinability of truth in the formal system", which "is credited… to Tarski (1933a) [1933 here]" (Kleene 1986a, p. 339; however, in his 1986, p. 127, Kleene follows Wang 1981 by writing that after his 1929 Gödel realized that "truth in number theory cannot be defined in number theory"). In the second case we read: "The undefinability theorem was added to the Polish monograph, with acknowledgment to Gödel, after it had gone to press" (Etchemendy 1988, p. 52). Before that, one is tempted to summarize the situation by saying that Tarski's Theorem of undefinability of truth was credited for Tarski and Gödel to one another. Let us review other accounts available in the literature.

Feferman (1984) devoted quite a few pages explicitly to our theme. Feferman quotes Wang (1981), who was the first one in writing that when Gödel considered first the re- lationship between the paradoxes, on the one hand, and truth and undefinability, on the other, he "realized that truth in number theory cannot be defined in number theory…" (1981, p. 654). Also, Feferman (p. 554) quotes the letter from Gödel to Burks, published in a book by von Neumann edited by Burks in 1966, according to which Gödel identified a theorem referred to by von Neumann as Gödel's, by writing: "the fact that a complete epistemological description of a language A cannot be given in the same language A, because the concept of truth of sentences of A cannot be defined in A", which is then pointed out as "the true reason for the existence of undecidable propositions in the formal systems containing arithmetic". Then Gödel adds that he did not formulate the theorem explicitly in his 1931a, but only in his 1934, and that: "The same theorem was proved by Tarski in his paper on the concept of truth published in 1933…".

According to that, Feferman offers us (p. 554) also the precise words by Gödel in 1934, concluding that the paradox of the liar "can be considered as a proof that 'false statement in B' cannot be expressed in B", although he emphasized the fact that the well-known footnote in which Gödel mentioned Tarski's (1933a, b) and 1944, as containing "a closer examination of this fact" was added only in 1965. Then Feferman goes on to a "reconstruction" of the precise sense in which for Gödel truth in number theory cannot be defined in number theory, as it can be regarded as giving place to a "heuristic argument" to construct undecidable number theoretical propositions. Finally, he adds a very illuminating explanation of the reason why Gödel eliminated any reference to objective mathematical truth in 1931: caution in view of the "philosophical prejudices" of the time. However, it is a pity that Gödel's own reconstruction in 1934 of this heuristic argument is not cited, together with the similar remarks in his letter to Zermelo of 1931 (although Feferman added in proofs a note pointing to this last fact, following Wang). We shall see below both versions of the same argument.

Feferman concludes that Gödel arrived at the undefinability of arithmetical truth in arithmetic in 1931, with which he recognized that "Gödel anticipated Tarski at least in establishing the undefinability of arithmetic truth within arithmetic" (p. 557), although "he did not state this as a result (only done so later, and independently, by Tarski in 1933)". However, something else about that would have to be said, especially by discussing the words "result" and "independently". The first, because if Gödel did not state the undefinability as a result somehow, it is difficult to understand how he could have anticipated Tarski in "establishing" it; the second, because there is a sense according to which it can be defended that Tarski did not state his famous theorem independently from Gödel. Let us leave these two questions for the following sections.

There is further interesting evidence on our theme in Dawson (1984a, b). Let us mention three points. Firstly, there is a reference to the Gödel-Bernays correspondence of 1931, after Gödel (1931a) appeared in January. As it happened, when Bernays read the famous paper, he "confessed his inability to see why a truth predicate could not be formally defined in number theory" (p. 4). Then Gödel insisted with his explanations, until Bernays admitted the point, which is judged by Dawson as very interesting, among other things, "because it furnishes independent evidence of Gödel's awareness of the formal undefinability of the notion of truth—a fact nowhere mentioned in (1931) [here 1931a]". Also, Dawson offers us some remarks on Gödel's correspondence with Zermelo, particularly by emphasizing Gödel's efforts "pointing out the impossibility of defining truth combinatorily within his system" (p. 5). Then he points out the several footnotes in Tarski (1935, 1956a, b), where he "makes clear his indebtedness to Gödel's methods" (p. 14, footnote 5). Finally, he also quotes a passage from Russell's *My philosophical development,* where Russell mentioned the fact that in his introduction to the *Tractatus* (1922) he already suggested that a higher order language makes possible to say things which cannot be said in a given language. This last remark, Dawson adds, may be of some interest to evaluate Gödel's originality in claiming "that by passing to successively higher order types one can obtain a transfinite sequence of formal systems such that the undecidable propositions constructed within each system are *decidable* in all subsequent systems" (p. 12).

I come now to Wang (1987), where Gödel's priority as for Tarski's theorem is strongly claimed. Wang begins by reproducing his 1981, so we can read again that already in the summer of 1930 Gödel realized that "truth in number theory cannot be defined in number theory" (p. 42). But after that Wang studied Carnap's unpublished personal records (now in the University of Pittsburgh), and he found there that on July 2, 1931 "G proved 'in the circle' that arithmetic truth is not definable (in arithmetic)". By recording Gödel's *proof,* Carnap wrote: "There are in arithmetized metalogic ordinary concepts which are not definable. This is proved by deriving a contradiction (after Richard) from supposing the concept available..." (p. 51). Then Hao Wang adds that the details of the proof are similar to the ones appearing in Gödel's letter to Zermelo of September 21, 1931 (published in Grattan-Guinness 1979), and later he offers us a clear treatment of that letter (pp. 91–2).

There we begin by reading that "the letter contains an exact proof that truth for a formal language is not definable in itself..., a theorem commonly attributed to Tarski, who proved it as a corollary to G's work only in 1933". Also, Wang offers us a reconstruction of the above named "heuristic argument" for the existence of undecidable propositions, which is here presented as "an alternative proof of his first incompleteness theorem". The proof is given this way: "Since *W* (the set of true sentences) cannot be represented (by a set sign of defining formula) in the given system S, but *B* (the set of propositions provable in S) can, and since *B* is a subset of *W,* *B* is a proper subset of *W.* Hence, there is some true proposition *A* that is not provable in S. Since *A* is true, its negation should not be provable in S either". Therefore, we can conclude that this proof of the first incompleteness theorem depended upon the previous proof that arithmetic truth is not definable in arithmetic. The reason why Gödel went on to another different proof (the only one finally appearing in 1931a) seems to be the twofold one which can be inferred from Gödel's further remarks: (i) this proof does not exhibit an actual undecidable proposition; (ii) it is intuitionistically objectionable (it seems due to the appeal to the set *W,* Wang adds).

11.3 Tarski's View

Tarski faced directly the problem of the priority of his ideas and results on truth in a number of places, which can be interpreted as a first sign that the matter was worth considering, and that things are not so clear as we are sometimes told. First of all,

Tarski tells us that most of the results of 1933 date from 1929, and that first reports of them were given in his 1931 and 1932 (1956a, p. 152, footnote). As for his Theorem I, in §5 of 1933, 1935 (see above, Sect. 11.1), Tarski wrote that the theorem, together with the sketch of the proof, was added after 1933 was already gone to press, then he goes on to the following remark about priority: "At the time the work was presented at the Warsaw Society of Sciences (21 March 1931), Gödel's article—so far as I know—had not yet appeared. In this place therefore I had originally expressed, instead of positive results, only certain suppositions in the same direction, which were based partly on my own investigations and partly on the short report, Gödel, K. (21), which had been published some months previously" (1956a, 247). The presentation to the Warsaw Society was his 1931 and Gödel (21) is here Gödel 1930a. Then we have to say that Tarski's results present in his Theorem I are not independent of Gödel's work. Besides, Gödel (1931a) appeared in January 1931, so it is prior to Tarski first report to the Warsaw Society in March. Therefore, for Tarski there must have been something in Gödel's incompleteness results which led him to the underlying idea of the undefinability of arithmetic truth in arithmetic, and this must have been the fact that incompleteness can be derived from that undefinability.

Yet later Tarski must have felt that things required further clarifications, as he decided to include a few "Historical Notes" at the end of the 1935 Postscript (1956a, 277–8). There he faces explicitly the problem of the dependence or independence of his methods and results. This is done by listing the particular points where complete independence is claimed, which includes the method of arithmetizing the metalanguage. However, this does not seem to include Tarski's use of Gödel's method of constructing undecidable sentences, for it is admittedly applied in Tarski's proof of Theorem I, which was included only after having read Gödel (1956a, 264; the same combination of this method with the results in theory of truth is claimed in Tarski (1939, p. 105). There is then only one thing left: the possible independence about undefinability of arithmetic truth. Here Tarski is completely explicit: "In the only place in which my work is connected with the ideas of Gödel—in the negative solution of the problem of the definition of truth for the case where the metalanguage is not richer than the language investigated—I have naturally expressly emphasized this fact (cf. p. 247, footnote)" (1956a, 278). And this seems to me to be equivalent to an explicit recognition of Gödel's priority for what it is usually known as "Tarski's Theorem".

It is however curious that in this recognition there is no further distinction between the two senses in which the expression is sometimes used (see Sect. 11.1 above). Be that as it may, the important thing for us here is noticing that in both senses there seem to be little doubt about the actual priority, although I think that further discussion is needed as for the precise sense in which undefinability of arithmetic truth in arithmetic is present in Theorem I (as mentioned above).[1] As for other "borderline" zones of priority between Tarski and Gödel, the only one I know is referred to Tarski's discovery of consistent systems which are not co-consistent, which took place before Gödel's (Tarski 1933a, p. 279), although with another terminology. However, even

[1] See the Addendum at the end of this paper, before the bibliographic references.

on that point, Tarski added: "Naturally it is not hereby claimed that I already knew then the results later obtained by Gödel or had even foreseen them".

Tarski still returned to this theme in his 1944. First, to reproduce the general argument according to which undecidability follows from the discovery that truth and provability are different concepts, given that every provable sentence is true (p. 372, footnote 17). Second, to point out how simply the fact that the theory of truth leading directly to Gödel's incompleteness theorems can be explained (pp. 372 f.):

> In deriving Gödel's result from the theory of truth we make an essential use of the fact that the definition of truth cannot be given in a metalanguage which is only as "rich" as the object-language...; however, in establishing this fact, a method of reasoning has been applied which is very closely related to that used (for the first time) by Gödel. It may be added that Gödel was clearly guided in his proof by certain intuitive considerations regarding the notion of truth, although this notion does not occur in the proof explicitly.

This is a further statement of Gödel's priority in the sense explained above, but it is very interesting because, in addition, Tarski seems to me to point out that: (i) undefinability of truth leads to undecidability; (ii) the method of constructing undecidable sentences was used first by Gödel; (iii) all that was already implicit in Gödel's incompleteness theorems.

11.4 Gödel's View, and an Unpublished Letter by Him

The ultimate origin of the difference between truth and provability mentioned by Gödel was already his conviction in the introduction to 1929 that "a proof of the unsolvability of a problem" cannot be excluded as soon as we state precisely certain formal means of inference (1929, p. 63), given that this remark seems to contain an anticipation of the existence of undecidable propositions (that is, propositions which cannot be proved nor disproved), then some anticipation of the underlying argument according to which it is precisely the undefinability of truth what leads to undecidability (see above). The same can be said as for Gödel (1931a, b, c), for there we can find the first explicit claim that undecidable propositions are possible: "... one can even give examples of propositions... that, while contentually true, are unprovable in the formal system..." (1931, p. 203). Moreover, there seems to exist some direct connection between this idea and the one of undefinability as in the same communication Gödel talked first about formal irrepresentability in a system (1931, p. 201):

> if in a consistent formal system *A* (say that of classical mathematics) a meaningful proposition p is provable with the help of the transfinite axioms, there follows from the consistency of *A* only that not-p is not formally provable *within* the system *A*. Nonetheless it remains conceivable that one could ascertain not-p through some sort of contentual (intuitionistic) considerations that are *not* formally representable in *A*.

However, this supposed connection is not clear to me, and it is not elaborated by Dreben and van Heijenoort 1986 nor by Dawson 1986 (although Dawson explicitly points out the mentioned link between Gödel (1929, 1931).

As for Gödel (1931a) a further interesting point in connection with Feferman 1984 is constituted by his remarks (p. 558) on Gödel's footnote 48, where Gödel pointed out the fact that "the formation of higher types can be continued into the transfinite..." as being the true reason for the incompleteness of all systems of mathematics, given that any undecidable proposition becomes decidable by adding higher types. Feferman writes that this can be analyzed, following Tarski, "by saying that the inductive definition of truth for a language can be given in an extension of that language to a next higher type". However, it occurs to me that for Tarski (1933a, b) this might be done only for languages of finite order, and that only from the Postscript added in 1935 onwards (then after Gödel 1934, although it was published only in 1965) the extension of his methods to infinite orders was admitted (see Sect. 11.1 above for my emphasizing of this historical fact). Besides, a more explicit connection between this note and truth has been stated by Kleene (1986a, b, p. 135):

> Implicit in this remark is that the adjunction of higher types to a formal system permits one to define the notion of truth for that system, then to show that all its provable sentences are true, and hence to decide the sentence shown in Theorem VI to be undecidable in the system.

Thus we see once again that it is undefinability of truth in a system (then its definability in a higher one) what led Gödel to undecidability. It is certainly curious that Gödel never published explicitly on the undefinability of truth, which is usually interpreted as meaning that when he saw Tarski (1933a, b) he thought the work needed had been satisfactorily done already. Feferman says (1984, p. 557) that there is no evidence that Gödel ever considered Tarski's problem (a set-theoretic definition of truth), but we must not forget that Tarski's theorem was added to Tarski (1933a, b) only after he knew some of Gödel's ideas. Besides, there *is* some evidence that Gödel was decided to include some definition of truth in the never published (and perhaps never written) 1931a-II. Thus Coffa 1987, in recognizing that the notion of truth is involved in Gödel's major theorems (p. 548), pointed out some passages from a letter from Gödel to Carnap of September 9, 1932, in which, after saying that in order to define "analytic" correctly we have to recognize no limitation in the range of variability of the function-variables (p. 553), Gödel added "On the basis of this thought I will give a definition of 'true' in the second part of my work..." (p. 569, footnote 5), which is presumably referring to the second part of his 1931a, and seems to point out to the theme of the need for higher types to define truth.

However, the first explicit statement about the relationship between the difference truth-provability and the existence of undecidable propositions appeared only in Gödel (1934) (printed in 1965). I already said above that Feferman (1984), in pointing out to this fact, wrote only that it contains Gödel's first statement of the undefinability of truth within the object language, in connection with the paradox of the liar. But as I also said, Feferman went then to reconstruct the "heuristic argument" leading from that fact to undecidability, by following former remarks by Wang (1981). And this, being entirely correct, may however lead some readers to think that the argument was not explicitly formulated by Gödel himself. We saw above that in his letter to Zermelo of 1931 the argument was completely developed, but in exactly the same way we can find it in 1934, where Gödel finishes §7 by saying (1934, p. 363):

> So we see that the class α of numbers of true formulas cannot be expressed by a propositional function of our system, whereas the class β of provable formulas can. Hence $\alpha \neq \beta$ and if we assume $\beta \subseteq \alpha$ (i.e., every provable formula is true) we have $\beta \subset \alpha$, i.e., there is a proposition *A* which is true but not provable. ~*A* then is not true and therefore not provable either, i.e., *A* is undecidable.

We can finish by wondering about Gödel's own view as for priority and independence in connection to Tarski's Theorem. I already cited Gödel's letter to Burks, published in 1966, where he wrote that undefinability of truth in the object language is the reason for the existence of undecidable propositions, and recognized that the theorem was implicit in his 1931a and explicit in his 1934, although without forgetting the fact that Tarski proved the theorem in his 1933. But in Gödel's *Nachlass* there is a copy of the original letter to Burks, dated on November 7, 1961, where the references to Tarski reads: "The same theorem was proved independently by Tarski in his book on the concept of truth". Then we can find also a second letter to Burks of February 6, 1964, where Gödel granted Burks his permission to print the former letter, but with the following significant changes related to our theme:

> Secondly the last line on p. 34 should be replaced by: "Tarski in his paper on the concept of truth published in 1933 in Act. Soc. Aci. Lit. Vars." The word "independently" was supposed to refer to my 1934 lectures. But since Tarski's paper actually came out before 1934, it is not necessary to mention this independence. On the other hand if "independently" were referred to my 1931 paper it would be incorrect, as appears from the footnote 1 on p. 247 in "Logic, Semantics, Metamathematics, Papers by A. Tarski, Oxford 1956."

If some explicit conclusion to these notes had to be stated, I think it *may* be defended that Tarski's Theorem is not really Tarski's, at least under the form of the undefinability of arithmetic truth. As we have seen, some of the reasons are the following: (i) it was known to Gödel at least from 1930; (ii) it was proved by Gödel in 1931, at least before Carnap (and perhaps before others); (iii) it was proved in the letter to Zermelo of 1931; (iv) Tarski arrived at the discovery of the theorem only after having read about Gödel's undecidability results; (v) even so, Tarski did not include *clearly* the undefinability of arithmetic truth in arithmetic in 1933—Theorem I—, but only in the Postscript from 1935; (vi) Tarski's dependence upon Gödel was recognized by himself and also by Gödel. I also think that almost the same can be said as for the general undefinibility of truth in the object language, although in that case perhaps the priority could be ultimately attributed to Russell's seminal ideas about languages of higher order, which were already implicit in his theory of logical types, and appeared more explicitly in his introduction to the *Tractatus* in 1922.

Addendum

As I write in the above paper, the expression "Tarski's theorem" is interpreted in different ways by different people; even logicians not always agree about that. In view of that very unsatisfactory situation, years ago I composed a sort of test of multiple choice, and sent it to a few very prestigious logicians or philosophers of logic, together with a copy of the paper. I received only three replies, but all of them, although extremely interesting in themselves, differed heavily. For obvious reasons it is advisable not the make their names public. The test follows.

Let the expression "Tarski's theorem" to mean this: "arithmetical truth is not definable in Arithmetic" (T'TA in the following). Also, some people choose that expression to mean Theorem I in Tarski (1935), §5 (1956a, p. 247), which says: (a) if a truth predicate "denoting a class of expressions, is defined in the metatheory, it will be possible to derive from it the negation of one of the sentences" described in the convention T; (b) if the class of all provable sentences of the metatheory is consistent, "it is impossible to construct an adequate definition of truth in the sense of convention T on the basis of the metatheory" (T'TI in the following).

Now, what is the relation between T'TA and T'TI? Could you please choose one or more possibilities among these?

So, T'TA is: (1) Identical to T'TI; (2) Contained in T'TI; (3) Presupposed by T'TI; (4) Implied by T'TI; (5) Nothing of the kind, and (6) T'TA was clearly stated only in the 1935 Postscript: "It is impossible to give an adequate definition of truth for a language in which the arithmetic of the natural numbers can be constructed, if the order of the metalanguage in which the investigations are carried out does not exceed the order of the language investigated" (1956a, p. 272).

References

Carnap, R. 1963 "Intellectual autobiography". In Schilpp, P. A. (ed), *The philosophy of Rudolf Carnap,* La Salle, III.: Open Court.

Coffa, A. 1987 "Carnap, Tarski and the search for truth". *Nous,* 21: 547–572. (Reprinted in 1991.)

Coffa, A. 1991 *The semantic tradition from Kant to Carnap,* Cambridge: Cambridge University Press.

Corcoran, J. 1983 "Editor's introduction to the revised edition". In Tarski 1983.

Dawson Jr., J. W. 1984 "Discussion on the foundation of mathematics". *Hist. Phil. Log.,* 5: 111-129.

Dawson Jr., J. W. 1984a "The reception of Gödel's incompleteness theorems". In Shanker 1988. (Originally in PSA, vol. 2.)

Dawson Jr., J. W. 1985 "Completing the Gödel-Zermelo correspondence". *Hist Math.,* 12: 66-70.

Dawson Jr., J. W. 1986 "Introductory note to *1931a, 1932e, f* and *g*". In Gödel 1986.

Dreben, B. and van Heijenoort, J. 1986 "Introductory note to *1929, 1930 and 1930a*". In Gödel 1986.

Etchemendy, J. 1988 "Tarski on truth and logical consequence", *Jrn. Sym. Log.,* 53: 51-79.

Feferman, S. 1984 "Kurt Gödel: conviction and caution". *Philosophia Naturalis,* 21: 546-562.

Gödel, K. 1929 "On the completeness of the calculus of logic". In Gödel 1986 (originally in German).

Gödel, K. 1930 "The completeness of the axioms of the functional calculus of logic". In Gödel 1986 (originally in German).
Gödel, K. 1930a "Some metamathematical results on completeness and consistency" (presented to the Vienna Academy of Sciences by H. Hahn on October 23). In Gödel 1986 (originally in German).
Gödel, K. 1931 "Discussion on providing a foundation for mathematics" (Königsberg Conference, September 1930). In Dawson 1984 and Gödel 1986 (originally in German).
Gödel, K. 1931a "On formally undecidable propositions of *Principia mathematica* and related systems I" (received for publication on November 17, 1930). In Gödel 1986 (originally in German).
Gödel, K. 1931b Letter to Zermelo (October 12). In Grattan-Guinness1979.
Gödel, K. 1934 "On undecidable propositions of formal mathematical systems". In Gödel 1986 (originally in German).
Gödel, K. 1986. *Collected Works. Volume I. Publications 1929–1936,* edited by S. Feferman *et al.*, New York and Oxford: Oxford University Press.
Grattan-Guinness, I. 1979 "In memoriam Kurt Gödel: his 1931 correspondence with Zermelo on his incompletability theorem". *Hist. Math.*, 6: 294-304.
Kleene, S. C. 1975 "The work of Kurt Gödel". In Shanker 1988.
Kleene, S. C. 1986 "Introductory note to *1930b, 1931* and *1932b".*In Gödel 1986.
Kleene, S. C. 1986a "Introductory note to *1934".* In Gödel 1986.
Parsons, C. 1974 "Informal axiomatization, formalization, and the concept of truth". In the author's *Mathematics and philosophy,* Ithaca, N. Y.: Cornell University Press, 1983. (Originally in *Synthese.*)
Shanker, S. G. 1988 (ed.) *Gödel's theorem in focus,* London: Croom Helm.
Tarski, A. 1931 "O pojeciu prawdy w odniesieniu do sformalizowanych nauk dedukcyjnych" (On the notion of truth in reference to formalized deductive sciences), *Ruch Filozoficzny,* 12 (1930-1): 210-11.
Tarski, A. 1932 "Der Wanrheitsbegriff in den Sprachen der deduktiven Disziplinen", *Akad. der Wiss. in Wien, Mathematisch-naturwissenschaftliche Klasse, Akademischer Anzeiger,* 69: 23–25
Tarski, A. 1933 *Pojecie prawdy w jezykach nauk dedukcyjnych* (On the concept of truth in languages of deductive systems), Warsaw.
Tarski, A. 1933a "Some observations on the concepts of ω-consistency and ω-completeness". In Tarski 1956, 1983 (originally in German).
Tarski, A. 1935 "Der Wahrheitsbegriff in den formalisierten Sprachen", *Studia Philosophica,* 1: 261–405 (revised translation of 1933, with a new Postscript).
Tarski, A. 1939 "On undecidable statements in enlarged systems of logic and the concept of truth". *Jrn. Symb. Log.*, 4: 105-112.
Tarski, A. 1944 "The semantic conception of truth and the foundations of semantics", *Phil. Phen. Res.*, 4: 341-76.
Tarski, A. 1956 *Logic, semantics, metamathematics: papers from 1923 to 1938,* translated into English and edited by J. H. Woodger, Oxford: Clarendon Press.
Tarski, A. 1956a "The concept of truth in formalized languages". In Tarski 1956, 1983 (translation of 1935).
Tarski, A. 1983 Second edition of Tarski 1956, edited and introduced by J. Corcoran, Indianapolis, Indiana: Hackett Publishing Company.
Wang, Hao 1974 *From mathematics to philosophy,* New York: Humanities Press.
Wang, Hao 1981 "Some facts about Kurt Gödel". *Jrn. Symb. Log.*, 46: 653–659. (Reprinted in 1987).
Wang, Hao 1987 *Reflections on Kurt Gödel,* Cambridge, Mas.: The MIT Press.

Chapter 12
Tarski's Intuitive Notion of Set

Abstract Tarski did research on set theory and also used set theory in many of his emblematic writings. Yet his notion of set from the philosophical viewpoint was almost unknown. By studying mostly the posthumously published evidence, his still unpublished materials, and the testimonies of some of his collaborators, I try to offer here a first, global picture of that intuitive notion, together with a philosophical interpretation of it. This is made by using several notions of universal languages as framework, and by taking into consideration the evolution of Tarski's thoughts about set theory and its relationship with logic and mathematics. As a result, his difficulties to reconcile nominalism and methodological Platonism are precisely located, described and much better understood.

Keywords A. Tarski · Set theory · Logic · Mathematics · Philosophy

Mathematics Subject Classification Primary 03-02 · Secondary 03Exx

> I represent this very rude kind of anti-Platonism, one thing which I could describe as materialism, or nominalism with some materialistic taint, and it is very difficult for a man to live his whole life with this philosophical attitude, especially if he is a mathematician, especially if for some reasons he has a hobby which is called set theory, and worse –very difficult. (Tarski, Chicago, 1965)

Tarski made important contributions to set theory, especially in the first years of his long and highly productive career. Also, it is usually accepted that set theory was the main instrument used by Tarski in his most significant contributions which had philosophical implications and presuppositions. In this connection we may mention the four definitions which are usually cited as supplying some sort of "conceptual analysis", both methodologically (the first one) and from the point of view of the results obtained (the rest): (i) definable sets of real numbers; (ii) truth; (iii) logical consequence, and (iv) logical notions. So, we could reasonably conclude that for Tarski set theory was reliable as a working instrument, then presumably as a conceptual ground. As we shall see, there are some signs that the reason for this preference

F. Rodríguez-Consuegra, *Russell, Gödel, Tarski*, Studies in Universal Logic,
https://doi.org/10.1007/978-3-032-08945-8_12

might have been its simple ontological structure as a theory, at least excluding the upper levels, the highest infinite.

Nevertheless, very little was known about Tarski's conception of set theory from the philosophical viewpoint, apart from some comments he made to his closest friends and collaborators, and the conjectures which could perhaps be made from a few scattered remarks here and there in his publications. Fortunately, the situation is very different when we look at the unpublished evidence. Tarski left unpublished quite a few pieces of his work over the years, especially expository work, with no new technical results, including lectures actually given and transcriptions of contributions to meetings. Some of those lectures, as well as a few letters, have been recently published, containing materials useful for knowing more about his notion of set. Yet there are still more materials which have remained unpublished so far, which are highly relevant for the issues involved. Among them, there are two contributions to meetings in 1965, and the records of Tarski's conversations with Carnap and Quine in the early forties.

In the following, the most relevant passages of all of those materials will be quoted and analyzed, to try to throw some light on Tarski's ideas about the notion of set, especially from the philosophical point of view, in an attempt to present a global, coherent position in the philosophy of mathematics. Like Russell and Gödel, Tarski was much more sincere about his philosophical tendencies when no publication was involved, so his deepest beliefs are much easier to discern from those materials. Thus, in this case the archival work has been once again the only effective way to try to understand Tarski's actual ideas. The resulting picture is a fascinating struggle between his nominalistic tendencies and his need to behave *as if* the mathematical entities with which he was working actually existed.

12.1 Universal Languages

It is usually believed, by following van Heijenoort, that after the semantic, or model-theoretic, revolution logic could no longer be seen as a universal language, especially in the Hintikka sense of a language regarded as a universal medium, and so that the old universality implicit in Frege and Russell was no longer possible. However, a few traces of universality can be discovered in Tarski's writings, and some of them are associated to set theory (ST), so we first need to discuss those traces. As we shall see, several senses of "universality" do appear in his work.

The expression "universal language" is used by Tarski in different senses. In the first sense, ordinary language is universal because it is unable to contain the fundamental difference between the object-language and the metalanguage; that is why ordinary language produces antinomies. In *Wahrheitsbegriff* (WB) Tarski called this type of universality semantic universality (Tarski 1983, p. 262), and in this sense the introduction of the truth predicate in the metalanguage does not transform it into a universal language. In this connection, Tarski's distinction between object-language and metalanguage, needed to develop important semantic concepts, is not compatible with a truly universal language, in the semantic sense. But there are other senses of the expression.

A second sense appears in a footnote to the same celebrated monograph on truth (Tarski 1983, p. 210, ft. 2):

> In order to give the following exposition a completely precise, concrete, and also sufficiently general form, it would suffice if we choose, as the object of investigation, the language of someone complete system of mathematical logic. Such a language can be regarded as a universal language, in the sense that all other formalized languages – apart from 'calligraphical' differences – are either fragments of it, or can be obtained from it or from its fragments by adding certain constants... (...) As such a language we could choose the language of the general theory of sets which will be discussed in §5...

The same idea appears later again, where it is said that a complete system of logic should contain all semantical categories occurring in the languages of deductive sciences: "this fact gives to the language mentioned a certain 'universal' character, and it is one of the factors to which logic owes its fundamental importance for the whole of deductive knowledge" (Tarski 1983, p. 220). Since this sort of universality makes possible for a language to embrace all other formalized languages of mathematical interest, I will call it mathematical universality. In this sense, a language which is mathematically universal should be capable to express the whole of mathematics.

In the last sentence of the quotation Tarski went so far as to consider the language of "the general theory of sets" as one of those universal languages in the mathematical sense, so we need to determine the relation between this general theory, as developed in §5 of WB, and what we know as standard ST today. A first difficulty arises from the fact that the last part of the quotation was added for the English translation of WB, appeared in 1956 (see de Rouilhan 1998), and that this can be related to the changes introduced in the original WB monograph by the famous infinitistic 1935 Postscript. In particular, this postscript has been pointed out by de Rouilhan as containing the abandonment of logic, regarded as a universal language, as opposed to logic as mere calculus. But as we shall see mathematical universality can survive that postscript.

A second difficulty is that the general theory of sets appearing in §5 of WB is a simplified form of the formal apparatus of *Principia Mathematica* (PM), but still containing types, and so it cannot be identified with standard ST. However, a few years later Tarski made clear that everything you do with type theory you can do with no types ST. Thus, in his lecture in Harvard from 1939–40 (Tarski 1995) Tarski said that when ST is correctly used, we can dispense with the complicated apparatus of type theory, and its variables of many kinds, and replace it by a single universe of discourse for ST, with just one type of individual variables. By individual variables he meant variables which range over individuals, not over properties, and other "abstract" entities, a requirement which could not be met by type theory, which uses variables of any type, so the PM type theory involves higher order logic. And this matched very well Tarski's well-known preference for nominalism for, although he was an as-if-realist (see below), he always chose, when possible, the simplest theory from the viewpoint of the ontological implications. In this way the theory might, presumably again, be somehow "reduced" to the material, actual world. (Yet see the final section of this paper.)

This 1939–40 idea deserves to be transcribed in its entirety in Tarski's own words, not only because it shows his viewpoint about the expressive power of the language of ST, but also because it will allow us to introduce a further sense of universality (Tarski 1995, pp. 154–65):

> It is not at all obvious at first glance that every mathematical discipline can be reduced to a formalized theory of the standard type. The crucial point here consists in carrying out such a reduction for the general theory of sets, since as we know from the work of Frege and his followers, and in particular from Whitehead and Russell's *Principia Mathematica*, the whole of mathematics can be formalized within set theory. Only a few years ago most logicians, under the influence of *Principia Mathematica*, were inclined to believe that the formalization of set theory required a much more complex apparatus; they thought, namely, that a so-called "theory of types" was required in one form or another, that one could not deal with a single universe of discourse, and that one had to use variables of infinitely many kinds. We know today that we can dispense with such a complicated apparatus, and that we can formalize the theory of sets within the standard type, like any other mathematical discipline. In order to carry this out, we need only one non-logical constant, namely the sign "∈" or the two-termed relation which holds between an element and a set to which this element belongs. Then the only concern lies in a careful selection of the axioms. They must be weak enough to escape the antinomies, but at the same time they must be strong enough to ensure, within our universe of discourse, the existence of sets which correspond to as large a class of sentential functions as possible...

Thus, since ST can be formalized with no use of type theory, and so it can be used to reduce the whole of mathematics, the language of ST fully deserves to be regarded as a *mathematically universal language*, our second sense of universality. For Tarski this universality should give ST a very special status in mathematics and metamathematics, as the ultimate tool for the foundational task, at least by the time when the lecture in Harvard was given. In this second sense our intuitive notion of set should be clear enough to serve as a true foundation for the whole of mathematics, as the whole of mathematics can be expressed in the language of ST.[1]

Besides, the quotation makes very clear that one of the main advantages of ST, once we dispense with types, is that we can work with just a single universe of discourse for the variables, so I think mathematical universality should be related to the problem of the fixed vs. variable universe of discourse, and also to Russell's old claim about the need for a totally unrestricted domain of the variable (for details, see my Rodríguez-Consuegra 1989, 2004a). I believe for philosophical reasons Tarski should be regarded as a universalist, so defending a fixed, unique universe of discourse when an important concept was in discussion, as it happens with his famous definitions of truth, logical consequence, and logical notions. As we have seen he is very

[1] Accordingly, de Rouilhan's thesis that the 1935 Postscript to WB supposed the abandonment of logic as a universal language seems to me to be inexact, at least in the broad sense in which Tarski regarded mathematical universality. Besides, due to the distinction between object-language and metalanguage, which is made in order to define truth without generating antinomies, WB was already lacking universality, at least semantic universality, before the postscript was added in 1935. Also, as we shall see below, mathematical universality was later to be linked to the universality of the universe of discourse, already from that postscript onwards.

clear in the Harvard lecture about this, at least for his semantic program in the 30's. A little reflection shows this to be very likely the case.

In the WB a universal domain is required, in this sense of universality, as what is defined is not truth in a particular sense, or in a particular model, but in a unique, universal sense, where the only model involved is the real, actual world. Thus, no relativization of truth is involved in the famous definition.[2] Such a goal is not always achieved in practice, since a simplified form of type theory is used across the monograph, and for Tarski type theory involved a renounce to a single universe of discourse. However, the 1935 Postscript adds the important modification to dispense with the theory of semantic categories. That theory was rather obscure, but it was somehow related to the simplified type theory used by Tarski (1983, p. 215, fn.). In this connection, the main change from the 1933 text to the 1935 postscript is usually described as the replacement of a theory of finite types by one of transfinite types.[3] Let us see what this actually involves.

For once, the role of the 1935 type theory seems to me not very clear. Thus, although in the languages considered variables for individuals and for classes and relations[4] are used (Tarski 1983, p. 268), and orders are kept (p. 269), Tarski introduces "variables of indefinite order which, so to speak, 'run through' all possible orders" (p. 271). He does so in a rather tentative way, by "reason of trials and other considerations" (*ibid.*), but he says to be "almost certain that we cannot restrict ourselves to the use of variables of definite order if we are to obtain languages which are actually superior to the previous languages in the abundance of the concepts which are expressible by their means" (pp. 270–271). And the latter was required to obtain the maximal level of generality for the definition of truth; thus, the concept of truth can now be defined for any language, no matter its order, as variables of transfinite order can now be used (p. 271). However, in this way the type theory used in former sections of WB, the simplified form of type theory, is transformed into a transfinite form of type theory, so one of the main differences with ST vanished. Thus, the new type theory was more closer to ST than ever, so to a true universal domain.

And this is confirmed by Tarski in a footnote, when he writes that from the languages considered "*it is but a step* [my italics, FRC] to languages of another kind which constitute a much more convenient and actually much more frequently applied apparatus for the development of logic and mathematics" (Tarski 1983, p. 272, fn.). And these latter languages are described as having a very simple structure, and using just variables of indefinite order, which all "belong to the same semantical category" (*ibid.*). This seems to me equivalent, or very close, to the introduction of a universal domain, especially because Tarski mentions Zermelo and his successors as having shown that "with a suitable choice of axioms it is possible to construct the theory of sets and the whole of classical mathematics on the basis provided by this language"

---

[2] See Simons (1987), §6, about this; according to Simons, this approach agrees well with Bolzano's.

[3] See Gómez-Torrente (2001), II, especially pp. 161–162.

[4] For the latter Tarski writes "variable sentence-forming functors" (Tarski 1983, p. 268), and these were formerly explained as referred to classes and relations (Tarski 1983, p. 213).

(*ibid.*). Agreed, the single step mentioned was not taken in the actual development of the postscript; this would have perhaps required a complete rewriting of the whole monograph. However, the path is clearly pointed out to the possibility to do so, i.e. to use ST as a more convenient and acceptable tool. Therefore, it seems clear enough that mathematical universality is now linked to a new formalization of ST where a single universe of discourse, a fixed domain, is used, exactly in the way described in the quotation from the Harvard lecture above.

The same universality takes place, I think, in the definition of logical consequence. It is true that in this connection some apparent counter-intuitive consequences may emerge, especially related to arguments concluding in the cardinality of certain existing things in the world, but none of them should be relevant for Tarski, as he usually accepted the axiom of infinity as a sort of logical axiom.[5] A single, global universe of discourse would make impossible that "different" definitions of logical consequence might result. (In a similar way, different concepts of number should result from the PM type theory without the clause of the unrestricted, universal domain of the variable, through the axiom of reducibility.) Or worse, without that single universe of discourse, the concept of logical consequence could not actually be defined at all: "...it is impossible to define the proper concept of consequence adequately whilst using exclusively the means admissible in the classical theory of types" (Tarski 1983, p. 413, fn. 2).[6] Thus, the renounce to type theory was required to reach the full universality needed to develop the semantic program Tarski had planned.[7]

Finally, in the 1966 logical notions lecture, it seems obvious that the universal, fixed domain is, again, the whole actual world, which is explicitly mentioned in the official definition (see the last section below).

Therefore, it should be noted that the fixed domain, the unique universe of discourse, could be considered as a truly philosophical requirement for Tarski's whole project of clarification of deductive sciences. This can be regarded as a third sense of "universality": the universality of our universe of discourse, so we can call this universality *domain universality*. As it was the case in Russell, for Tarski the

---

[5] See Gómez-Torrente (1996) and Bays (2001) for opposite views about this issue, all of them raised by the seminal Etchemendy (1990). Mancosu (2005a) is a recent defense of the fixed domain interpretation, based on unpublished evidence.

[6] It could perhaps be objected that, since what Tarski is mentioning here is just classical type theory, he still might believe that the same definition of logical consequence was possible in the simplified form of type theory actually used by him in former publications of the 30's. But this is simply not the case. Tarski made totally clear that he could not give the definition in former writings precisely because he "wished to avoid any means of construction which went beyond the theory of logical types in any of its classical forms" (Tarski 1983, p. 413, fn. 2). And the "classical" form used in those writings was precisely his well-known, simplified form of type theory, the one dispensing with ramification and the axiom of reducibility; see papers VI, VIII (WB) and IX of (Tarski 1983). Therefore, even with the usual simplification type theory made the fundamental definition impossible.

[7] In the 50's Tarski used a standard, model-theoretic definition of logical consequence, where the universe of discourse is relative to a model, so domain universality is lost. Yet the entities involved would still belong to the set-theoretical universe.

universe of discourse should be totally unrestricted, except for working with "partial" theories, where we are trying just to play with the notion of interpretation of one theory into another.[8]

Yet domain universality was introduced at a price. As Tarski said in the Paris 1935 congress, in his philosophical talk on the establishment of scientific semantics, the semantic concepts should be defined "in terms of the usual concepts of the metalanguage and are thus reduced to purely logical concepts, the concepts of the morphology of language", and this procedure "proves to be closely connected with the theory of logical types" (Tarski 1983, p. 406). But as we have seen, the only way to obtain domain universality involved an actual renounce to type theory, so Tarski's original goal to avoid semantics to be "an independent deductive theory" (Tarski 1983, p. 405) were not to be reached. And this was important for him, since by that time he believed that when semantic concepts cannot be reduced to logical (or physical) concepts we cannot proceed in "harmony with the postulates of the unity of science and of physicalism" (Tarski 1983, p. 406). Therefore, Tarski could not show the semantic concepts to be ultimately "logical" because he preferred mathematical universality to be linked to domain universality. (See last section below, for the question whether or not the very set theoretical notions are logical.)

However, there is still another sort of universality, a fourth one. In the last letter to Neurath of 1936, Tarski wrote that a universal language is not enough, because it should force us to introduce the usual semantic concepts (Haller and Tarski 1992, p. 27), which I take to mean that this is impossible because it would produce antinomies, in not distinguishing between object-language and metalanguage, so universality here is the usual one: semantic universality. But immediately he goes on to say that if that problem was taken aside in the development of an actual science, some might then think that a single universal language would suffice, but this would be still incorrect for this reason (p. 26):

> To pursue actual sciences, something like physics, one must have available an extended mathematical apparatus; now we know however, that for every language (therefore also for the presumed 'universal language') one can give entirely elementary number-theoretic concepts, respectively sentences, which in this language cannot be made precise, respectively cannot be proved.

Therefore, no powerful scientific language can be universal. For one thing, thanks to the Löwenheim-Skolem theorem (LS) every consistent theory has a model in arithmetic; but then, by the Gödel incompleteness result, arithmetic has to contain

[8] Domain universality is not contradictory with Tarski's well-known relativism about what notions should be regarded as logical and as non-logical. Since his 1933 definition of truth (originally earlier), which is equally valid for mathematical and for factual truths, he always defended the impossibility to draw a clear line between logical and factual notions. This line is an arbitrary one trying to split all notions into two kinds, but all of them belong to the same, unique world. For the same reason, his well-known relativism concerning the line of separation between logic and mathematics seems to proceed from his assumption that the important thing is the unique world underlying both of them, so the fundamental property of a language trying to describe this world is its mathematical universality.

sentences which are undecidable. If this is so, then there is a further sense of universality: to be really universal a language should be complete. We can call this fourth sense of universality *completeness universality*. So, it seems that what is involved is that only very elementary languages (e.g., sentential logic) should be regarded as universal languages in this sense. And what about ST? According to Tarski this could be regarded as a mathematically universal language, but ST is also incomplete since Gödel 1931, so perhaps what Tarski had in mind is that if ST is regarded as a universal language, then it fails because it is incomplete: is has to contain undecidable sentences.

Thus, although the language of ST can be mathematically universal, it cannot be completeness universal, but this is no big deal, as no mathematically interesting language can be completeness universal. Also, ST can be domain universal, though not semantically universal, but this again is no big deal, because only ordinary language, or any language incapable of supporting the distinction between object-language and metalanguage, can be semantically universal, and semantic universality is the kind of universality generating antinomies.

Domain universality and mathematical universality fit each other very well from the philosophical viewpoint. If there is just one world, the real, actual world, then the language trying to mathematically describe this world has to have this world, a world of individual objects, as the unique universe of discourse, so this language, either the language of set theory or a similarly powerful one,[9] should be regarded as a universal language.

It has been written by Niniluoto (1999, p. 18) that Tarski's position concerning the universality of language was a hybrid between the classic conception of language as a universal medium (one language, one world) and the model- theoretic conception of language as calculus (many worlds, many languages), as he thought there is just one world but many languages. In view of the many senses of "universality" that we have found in Tarski, this should be qualified in the sense that, if we accept the universality of ST as a language for all deductive sciences, its mathematical universality, especially when this is linked to domain universality, then there could be just one language as well, and so Tarski should be considered rather on the universalist side.[10]

12.2 Conversations with Carnap

In the academic year 1940–41 Tarski was in Harvard, together with Carnap, Quine, Russell and Hempel. A "Logic Group" was constituted, in order to hold regular meetings, and more private conversations took also place, mostly between Carnap, Tarski and Quine. Carnap took notes of those discussions, which are now in the Carnap papers in the Pittsburgh University. Tarski's contributions to those discussions

[9] See below for algebra as a possible replacement candidate.

[10] See Feferman (2004a, b) for more details about this controversy.

are very interesting from the philosophical viewpoint, so in this section some of the things Tarski said are quoted and discussed.[11]

The main issue of the discussion was the problem of a possible language to serve as a ground for the unity of science, so the logical-factual distinction, the nominalism vs. realism dispute, and the merits of a finitistic language were questions which emerged very often. Tarski systematically defended the ultimate need for a finitistic language, his well-known position against any analytic-synthetic distinction, and his nominalistic tendencies. As he fixed his global position in a very brief, threefold statement we can start by quoting it (Jan. 10, 1941):

I understand at bottom only a language that fulfills the following conditions:

1. *Finite* number of individuals.
2. *Realistic* (Kotarbinski): The individuals are physical things.
3. *Non-Platonic*: Only variables for individuals (things) occur, not for universals (classes etc.).

I think we have to take this very seriously. First, we can wonder what does it mean to say that he understood just finitistic,[12] first order languages. I believe he must be thinking of understanding as something similar to "visualize", or to "imagine", in the same way we imagine or visualize physical entities, situations or processes. But second, we can also wonder, in view of this "practical" limitation in comprehension, how did he manage to "understand" higher languages, i.e., languages of higher order involving very complex variables. Fortunately, we have Tarski's reply to this question, just in Tarski's comments following the quotation above.

He says that he understood any other language in the same way he understood classical mathematics: as a calculus. And by that he offers the following explanation: "With whatever higher 'Platonic' statements [unreadable?] in a discussion, I interpret them to myself as statements that a fixed sentence is derivable (or derived) from certain other sentences." Therefore, higher order languages, and variables, were comprehensible to him just in a proof-theoretic way, and by this he meant that no deep comprehension was involved, in the visual, intuitive way, as it happens usually when we are in contact with physical objects. So, we can perhaps call "physical intuition" to what was underlying the model of his actual comprehension of mathematical entities and statements. As we shall see below, there are other hints to support this interpretation.

As for the fact that Tarski mentioned Kotarbinski, it should be remembered that he dedicated WB to him, and very often he mentioned him as his favorite philosopher,

---

[11] Greg Frost-Arnold is writing a Ph. D. dissertation about these records, entitled "Carnap, Tarski, and Quine's Year Together: Logic, Mathematics, and Science", and also planning a book to make Carnap's full notes public. He was so kind as to sending me his English translation of the notes, which I am using here. Also, Mancosu (2005b) is a first attempt of exposing the main lines of discussion between Tarski, Carnap and Quine in those conversations.

[12] Tarski later relaxed the first requirement, about the finiteness of the domain, to "we make no assumptions about the cardinality of the domain." Also, some acceptance of number theory as applied to physical objects seems to emerge here and there.

and his philosophical position, sometimes referred as "reism", as the closest to his own. Also, the fact that Tarski allowed the editors of his *Collected Papers* (Tarski 1986a) to include a paper by Kotarbinski, which was partially translated by Tarski himself, has been mentioned by Corcoran as a sign that he considered Kotarbinski's paper as a sort of summary of his own position in philosophy (see Corcoran 1991).

From this starting-point Tarski was consistently defending his nominalistic position, but there are a few more remarks specifically relevant to his notion of set. Let us have a detailed look at them.

First, there is a short remark where Skolem is explicitly mentioned. There Tarski was trying to state his main philosophical requirements for a true, finitistic viewpoint, avoiding Platonism. Then he adds that number theory has to be excluded from a language of science because by Skolem we know that all mathematics have a model in number theory (Jan. 10, 1941):

> Why is even elementary arithmetic, with countable domain, excluded? Because, following Skolem, all of classical mathematics can be represented through a countable model, so it can be expressed in elementary arithmetic, e.g. when one takes $\in$ as a certain relation between natural numbers.

Let us first suppose this referred to Skolem's celebrated 1933 result that elementary arithmetic can be shown, in an effective way, to have non-standard models. So if by LS every mathematical theory can be "reduced" (in no effective a way though) to number theory, and number theory have non-standard models, i.e. exotic, freaky models which have very little to do with numbers, then our intuitive notions of mathematical "objects" are totally jeopardized, in the sense that we might not know what are we talking about when talking about them.[13]

Also, what has this to do with ST? Well, number theory is reducible to ST as a mathematically universal language (see above), and ST is also vulnerable to LS. Thus, if ST is consistent, as we believe it is, then it has a model, then it has a countable model (a model in number theory, in the domain of natural numbers), so by Skolem it has non-standard models as well, then ST might not be reliable enough from the foundational viewpoint. The axioms of ST would then have models very far from our intuitive notion of set, the model we think we have in mind when we talk about sets, and when we use and apply ST, so ST would not be safer than arithmetic, and might not be a good basis for foundational tasks. Thus, the assumption that Tarski was referring to Skolem 1933 allows us to somehow understand Tarski's assertion that elementary arithmetic should be "excluded".

Now let us suppose that Tarski was referring just to LS, so not thinking of Skolem's non-standard models of arithmetic. Then all consistent mathematical theories will have a countable model in arithmetic. This includes ST, in whose case membership is interpreted as a relation among natural numbers. Then the only reason to "exclude" arithmetic might be that in this way number theory would allow the whole of mathematics, so including lots of Platonic entities, to be accepted in the language

---

[13] In the context of Tarski's physicalist attitude and rejection of classes, talk about mathematical objects should be always understood in the reductive sense; see below on Tarski's Platonistic reductionism.

of science which Tarski, Carnap and Quine were looking for at that time. Well, both interpretations seem to be possible. I like the first one best, mostly because of its philosophical interest, although I admit the second to be much simpler, and perhaps the right one.

At any rate, something is clear: Skolem's well-known "relativism" of set theoretic notions is independent of his construction of non-standard models for arithmetic, as it was formulated long before. As Skolem wrote already in 1923 (I quote after Wang 1996, p. 127):

> The most important result above is that set-theoretic notions are relative. I had already communicated it to F. Bernstein in Göttingen in the winter of 1915–16. (...) I believed that it was so clear that axiomatization in terms of sets was not a satisfactory ultimate foundation of mathematics that mathematicians would, for the most part, not be very much concerned with it. But in recent times I have seen to my surprise that so many mathematicians think that these axioms of set theory provide the ideal foundation for mathematics; therefore it seemed to me that the time had come to publish a critique.

This was of course applied to the continuum hypothesis (CH) and similar problems (*ibid.*):

> Since Zermelo's axioms do not determine the domain B [the model for them], it is very improbable that all cardinality problems are decidable by means of these axioms. For example, it is quite probable that what is called the continuum problem is not solvable at all on this basis; nothing need be decided about it.

So Skolem was probably convinced that when we talk in general of "set" or "set theory" we do not know what we really mean, or even that by those expressions nothing clear could actually be meant at all.[14]

It is then likely that those "limitative" results (Skolem 1933 and Gödel 1931) helped Tarski to leave his work in ST (see below) after finishing with his "semantic program" in the 30's. This can be now illustrated by some evidence in Tarski's conversations with Carnap, where we find a very surprising paragraph. They were looking for a global, unified language for science, in the spirit of the old Vienna Circle, so Carnap asked Tarski if that language should be constructed with or without types. Then Tarski replied (March 6, 1940):

> Perhaps something completely different will develop. It would be a wish and perhaps a guess, that the whole general set theory [die ganze allgemeine Mengenlehre], as beautiful as it is, will disappear in the future. Platonism begins with the higher levels. The tendencies

[14] For an excellent presentation and analysis of Skolem's set theoretic relativism, see Jané (2001). By the way, this paper mentions, as reported by Skolem, that in 1958 Tarski made an objection to Skolem that his insistence on countable models could be reversed by using the upward LS, so presenting the reductionistic flavor towards uncountable ones instead. Jané shows how Skolem dispensed very quickly with this criticism, but for us it requires of some discussion. After all, if Tarski was somehow vulnerable to Skolem's relativism, why he did not accept it and was rather apparently objecting to it? I think that Tarski's argument could be seen not as an attack to Skolem's "partial" relativism, but rather as an endorsement that this relativism can be constructed both ways, so presented as a still deepest form of relativism. A different take on this criticism will be considered at the end of the paper.

of Chwistek and others ("Nominalism") to talk only about designatable things [Bezeichenbarem] are healthy. The only problem is finding a good execution. Perhaps roughly of this kind: in the first language, natural numbers as individuals, as in Language I, but perhaps with unrestricted operators; in the second language, individuals that are identical with or correspond to propositional functions in the first language, thus to the properties of natural numbers expressible in the first language; in the third language, the properties expressible in the second language are taken as individuals, and so forth. So in each language one has only variables for individuals, but nevertheless they cover entities of different levels.

First, it is surprising that for Tarski the whole general ST might disappear in the future! The expression he uses is the same we already discussed in the first section above, so he presumably was referring to the same, the general theory of sets, with types, used in §5 of WB. The further reference to "higher levels" would match this interpretation, as those levels were used in type theory. However, it is hard to see why Tarski might have thought in 1940 that type theory could "disappear" in the future, since the typical theory of types from PM was hardly used by that time, not even in the simplified form of §5 of WB. Besides, as we also saw above, for Tarski type theory could be fully dispensed with and replaced by no type ST, so obtaining the same mathematical universality in a simplest way, plus domain universality.[15] Therefore, it is likely he was thinking just of some sort of ST, which would explain that he found it "beautiful", an adjective he probably would not use to describe type theory. If so, this would be compatible with his increasing doubts about ST, as we shall see in the following.[16]

As for the possible replacement for general ST, at first I thought he was perhaps thinking of something much more general, as category theory, where membership is defined, then eliminated, so ST could be reduced to category theory. But category theory developed from 1945 onwards, and it seems that Tarski did not like category theory anyway, probably for not allowing the sort of individual variables he was looking for. Thus, perhaps he was thinking of some sort of a very general algebra and foresaw some ways to use it as a main tool for the foundations of mathematics. If this was so, that could help understanding the 1940's "algebraic shift" which has been mentioned as a turning point in Tarski's development.

Tarski's algebraic shift of 1937–1940 is well documented in Givant (1991, 1999). He probably felt that some sort of algebra could serve as a new foundation for mathematics, returning so to the point of view of Löwenheim and Schröder. In particular, Tarski showed that his algebra of binary relations was able to provide with "a suitable framework for formalizing all of mathematics" (Givant 1991, p. 24; see Givant 1991, pp. 53 ff.). This might have been seen by Tarski as a solution of all the problems associated to PM type theory, and also to the ultimate indeterminacy of ST.

---

[15] As I pointed out in the first section above, although Tarski originally tried to use type theory in his semantical program, he was somehow forced to dispense with it in favor of ST, so as long as his own writings were concerned, the "disappearance" of type theory had already taken place. He did not like ST either, but ST was more convenient for certain purposes. See the last section below for more details about this issue.

[16] See Mancosu (2005b), especially §2, for more details about Tarski's ideas to dispense with type theory in his conversations with Carnap.

From the ontological viewpoint, algebra was certainly better suited for a nominalist: "Even algebraic symbols were taken to represent particulars, particular numbers or particular quantities" (Hintikka 2004, p. 363). Algebra, like ST, is not completeness universal nor semantically universal, but after Tarski's elaboration may be mathematically universal and domain universal. The main difference from the ontological viewpoint may be that the individuals constituting the algebraic domain might be seen as more akin to what Tarski had in mind when requiring genuine individuals as the raw material for the individual variables to range over. If this interpretation is correct, we would be before a very important, technical option, which may have been inspired on clear philosophical reasons.[17] At any rate, Tarski continued to use ST when it was convenient to him.

Also, in the quotation he talks about general ST as a beautiful thing, and this seems much easier to explain, as he always loved classical mathematics, and considered mathematics not only as a science but also as an art. As Corcoran has recently written (in correspondence with me):

> Tarski loved classical mathematics including Euclidean geometry, number theory, algebra, and analysis – he never held back on anything classical out of philosophical scruples. He hated people saying half-baked skeptical things about classical mathematics – this included Wittgenstein (at the top of the list, cf. Feferman account), the intuitionists, the logicists and the formalists. Tarski was a Platonistic reductionist –he used whatever Platonic entities that were needed for classical mathematics and its metatheory– but he always imagined that these apparently Platonic entities were really somehow reducible to physical or mental things.

Next, he praises nominalism from his teachers, which shows that his philosophical tendencies were a very strong influence on him. This could go even so far as to somehow have compromised some of his technical options, as it can be clearly seen in the quotation. This does not seem to be totally compatible with Corcoran's opinion above, but we must not forget that we are talking of unpublished evidence, taken from records of private conversations. As we shall see below, in similar contexts Tarski felt much freer to express his deepest philosophical beliefs.

Finally, in the quotation Tarski seems to presuppose that having only individual variables dealing with entities of different levels would be a good accomplishment. This shows that his preference for the least ontologically exuberant system was a very early one. Also, this can be clearly connected with the above-mentioned Harvard lecture's preference for ST over type theory on philosophical grounds. (Yet things may not be so simple, as we shall see in the last section below.)

As for Corcoran's suggestion to call Tarski's position "Platonistic reductionism", it is a very interesting one, so I shall use it in the following. As a working logician and mathematician Tarski behaved as if the entities which he used actually existed, though ultimately he admitted just individual, physical objects, so he could be called an as-if-realist also. Similar positions have received similar names: methodological

---

[17] Tarski's big output after the algebraic shift is well represented by his work on cylindric algebras, which may be seen as the algebraic way to do first order logic, and on ST without variables, which might perhaps be interpreted as a way to dispense with the problematic range of the ontologically unrestricted variables of ST.

realism, working realism, or internal Platonism,[18] but I think Platonistic reductionism is more informative, for it adds the reductive element. Besides, it matches very well the physicalist tendencies that can be found in Tarski's work, and physicalism is clearly a reductivist position.[19]

After all, it was in the spirit of the old logicism, and also of the old logical empiricism, to present most of the entities under consideration as a sort of "logical constructions" in terms of more solid, reliable entities (see my Rodríguez-Consuegra 1991) for details about Russell's constructive definitions). So Platonistic reductionism could be seen as Tarski's own version of those old tendencies. Yet the clear empiricist tendency always present in Tarski's philosophical approach should not be interpreted as if Tarski was somehow returning to the old logicist masters: his sort of reductionism should even be regarded as a pioneering work towards an empiricist philosophy of mathematics, or even towards a set theoretic naturalistic position as it is well-known today (see for instance Maddy (1996). Finally, reducibility to physical or mental entities should be understood, I think, in the sense that all mental entities are ultimately also reducible to physical ones, as seems to be clear in Tarski's quotation above concerning Kotarbinski.

12.3 Is Set Theory a Wasting of Time?

Let us return to ST. As we have seen, by the end of the 30's ST was already infected by several limitations which clearly affected its conceptual basis. Because of the Gödel limitations, ST could not really be regarded as a universal language in the sense of completeness universality, and due to LS and to Skolem 1933 there could not possibly be a clear model for it. But the situation was to become still worse: in 1939 Gödel showed the axiom of choice (AC) and CH to be consistent with the standard axioms of ST. A first, initial implication of this was some talk about non-Cantorian ST, in case it could also be proved that the denial of CH was also consistent with the Zermelo-Fraenkel theory with the axiom of choice (ZFC) (which had to wait until Cohen 1963), so the Gödel result could be seen as partially confirming the Skolem suspicious attitude towards the indeterminacy of ZFC.[20]

---

[18] See Jané (2004), and the references cited there to Shapiro and Ferreirós.

[19] See Frost-Arnold (2004) though. He has showed quite convincingly that physicalism was not Tarski's main motivation in WB, but solving the semantic antinomies. Frost-Arnold's criticism of the classical Field (1972) paper is also convincing because that paper was based mostly on a unique mention by Tarski of the word "physicalist". Yet there is another mention of the same tendency in Tarski's contribution to the1965 Chicago meeting, although it seems it was made in a rather casual way.

[20] On the other hand, Gödel's 1939 result can also be seen as increasing the reliability of ST, as compared with type theory. Thus, those who formerly adopted type theory for fear that further contradictions appeared in ST, could then regard ST as a more reliable tool: "These consistency results undermined any remaining cautionary reason to adopt TT [type-theory]" (Ferreirós 1999, p. 198). However, Tarski had other reasons, not only technical but also philosophical, to dislike ST.

All in all, this could help understanding what Tarski said to Carnap in 1941 that "general" ST might disappear in the future. As we saw above, this and similar results may have caused Tarski to abandon his technical work on ST and also help to explain his algebraic shift. This is confirmed by Givant's recent letter to me:

> Concerning the many of the open questions in set theory, his attitude was rather negative. Most of these statements had been proved to be independent of the usual axioms of set theory, and seemed to be unresolvable by reasonable new axioms that would be accepted as "self-evident" by the majority of the mathematical community. He believed that much of the current research into the foundations of set theory —the results concerning the truth or falsity of this or that statement in various set-theoretical models— would eventually hold little interest for most mathematicians and would even be viewed by future generations in much the way that current philosophers view certain religious disputations of medieval scholastics. He thought that people would gradually stop working in set theory because there were too many unanswerable questions.

Yet Tarski continued to work on ST when interesting problems appeared. And this can be easily interpreted as one more evidence of the tension between his philosophical attitude and his research work.[21] As Givant adds in the mentioned letter:

> I would like to distinguish between Tarski's personal philosophical point of view and his actual practice in his mathematical work. He was not inhibited in his use of transfinite, or infinitistic, methods, and this applied to all aspects of his work, not only to his work in set theory. In fact, I believe he found certain more finististic aspects of logic, such as recursion theory, distasteful. Most importantly, he was very much concerned with problems he thought would be of interest to a broad spectrum of mathematicians —not just to logicians. (...) Tarski did regret that he had spent so much of his life working in set theory. I am not sure when this regret was first felt, but already by 1930 he had begun to move away from set theory. Still, when interesting set-theoretical problems occurred to him, even late in life, he would follow through on them. His papers with Erdös and with Keisler are just two examples. So it seems that disillusionment with set theory did not carry with it, as a corollary, total rejection of this domain. His intellectual interests were always the dominant force, not his philosophical beliefs.

Thus, it seems that the possible impact of CH independence results on Tarski is a matter to be somehow investigated. However, there was no apparent reaction to these results in the published work by Tarski known to me. But if we study the unpublished evidence things are very different.

12.4 The Continuum Hypothesis

The first reference known to me is the one appearing in a letter to Gödel from 1946 (Wolenski and Köhler 1999, pp. 271–2; I quote after Sinaceur 2000, p. 13). In the letter Tarski is announcing a part of what he was planning to say in the Princeton conference (Dec. 1946), and we read:

---

[21] That unresolved tension was already pointed out by Mostowski (1967b).

> As regards the question in which you are interested [absolute provability, definability, etc.], I don't think that I can do anything else [in the Princeton Bicentennial talk] but to emphasize the fundamental difference between all the undecidable statements known at present in elementary number theory on the one hand and some undecidable statements (like continuum hypothesis) in analysis and set theory; the statements of the first kind being clearly undecidable in a relative sense while those of the second seem to be undecidable in some absolute sense.

The first thing to notice is that he thinks CH to be undecidable, then independent from ZFC, so he somehow foresaw Cohen's result in 1963. Also, he thought CH to be absolutely undecidable, not just undecidable in relation to a certain axiomatization of ST. Yet in the actual talk Tarski said very little on CH, and after just mentioning the results of Gödel on CH, he added (Sinaceur 2000, p. 26):

> In discussing the relations between set theory itself and logic I shall begin by saying that I believe that the set-theorist may expect much from the formal logician. I believe that certain problems of set theory may actually be independent of the axioms of set theory and may be shown to be so independent by formal logical means.

So, there is no explicit mention of his actual belief that CH might be absolutely undecidable according to the letter to Gödel. Also, he does not mention CH explicitly, and the example he offers after the quote is not referred to CH but to another technical problem (the Suslin one). Finally, it seems to me he was thinking of CH as well, in writing about the independence of ZFC. So, we seem to be once again before one more example of the "conviction and caution" attitude, so dear to Gödel, but also very well-known to Tarski.

All in all, Tarski seemed to believe that CH might result to be undecidable, yet at the same time, in not making his belief public in the actual talk, he might believe that perhaps CH could be decided somehow in the future. This seem not to be compatible with what he actually wrote in the letter to Gödel (Gödel believed CH to be false!), where he talks about absolute undecidability, but even so he wrote just that this seemed to be the case. At any rate, Tarski did not share Gödel's intuitions about a future, when stronger axioms for ST might be found that would bestow an absolute truth value on CH (see below, a further quotation from Givant's letter).

So, in the 40's Tarski must have thought of ST as something not very reliable, in the sense of a good basis for his foundational projects. However, although ST could not be a universal language in the semantical sense, nor in the completeness sense, it was still capable to be a universal language in the mathematical sense: all mathematics can be reduced to ST, on pain, of course, of having to face the question of the non-standard models, and so the question of the low profile characterization of its "intended" model.

12.5 Are There Large Cardinals?

Large cardinals are those which cannot be constructed, i.e. their existence proved, by using the usual ZFC axioms. Inaccessible cardinals (IC) are among large cardinals (though there are many more classes of them) and are so exorbitantly large that the hypothesis of their existence can be used as a test about the ontological implications of ST and about the reliability of set-theoretic intuition. Tarski made a few contributions to the theory of those cardinals in the 30's and 40's, but did not return to the subject until the 60's. As we shall see, his latter attitude about ST was more skeptical, so this subject is useful to understand his philosophical evolution, especially because Tarski did not believe in large cardinals (John Addison 1996, personal communication).

In Tarski (1938, 1939) there seems to be a rather bold attitude about the existence of IC, to the point that in them he "formalized the existence of arbitrary many inaccessible cardinals as an axiom, phrased in such a way that most of the axioms of set theory including the Axiom of Choice are derivable" (Kanamori 1997, pp. 20–21). Curiously enough, this was done little after Zermelo's ontologically enthusiastic second order formalization of ST in 1930.

In his 1943 paper with Erdös and Tarski (1943) certain properties of cardinals are described which imply inaccessibility, and the difficulties to solve the problems involved are finally suggested not to depend on the nature of IC, but rather on the methods used, so (I quote after Kanamori 1997, pp. 70–71):

> It is quite possible that a complete solution of those problems would require new axioms which would differ considerably in their character not only from the usual axioms of set theory, but also from those hypotheses whose inclusion among the axioms has previously been discussed in the literature and mentioned previously in this paper (e.g., the existential axioms which secure the existence of inaccessible numbers, or from hypotheses like that of Cantor which establish arithmetical relations between the cardinal numbers).

As we shall see in a moment, the belief in the possible need of those "new axioms" was rejected when Tarski returned to the subject many years later.

Thus, in his paper to the Stanford 1960 conference we find something else which is useful for understanding Tarski's deep attitude towards ST, and the way in which he may have taken philosophically inspired options in his technical work. The paper (Tarski et al. 1962) has to do with certain very large cardinals; in particular, with the problem whether new set-theoretical axioms are needed to solve some problems related to IC. And this is said almost exactly with the same words than the ones used in the former quotation from 1943, except that instead of "It is quite possible…" we find "It seemed even plausible that…" (Tarski et al. 1962, p. 125).

The expression "new axioms" are probably somehow referring to the "stronger" axioms Gödel had said that, when they are added to the usual axioms of ST, they could solve, for instance, CH and similar problems. Tarski does not mention Gödel explicitly though, and says that those axioms would differ from general CH or existential hypotheses assuring the existence of IC, but it seems to me quite clear that he was thinking of the ontological problems raised by Gödel (see below).

The paper uses infinitary languages and the compactness theorem for them (as introduced in Tarski 1958), and is based on results obtained by Hanf, one of Tarski's students, "as a solution of a problem suggested by the author" (Tarski et al. 1962, p. 125), so it is clear that Tarski seemed to have been looking for something in that direction. This is very interesting, as the main claim of the paper is that *those strong axioms are not needed*, and that the cardinals involved in those problems behave "like all the accessible cardinals" (*ibid.*). This is not a denial of the existence of large cardinals, which had pleased Tarski indeed, as he did not believe in them, but at the end of the paper we can find some considerations with a clear philosophical flavor, precisely in that negative direction. Thus, we may be here at one place, probably unique in his publications, where Tarski's philosophical nominalistic attitude towards ST seems to be closely related to his technical options and results.

According to Tarski et al. (1962, p. 134):

> The belief in the existence of inaccessible cardinals $> \omega$ (and even of arbitrarily large cardinals of this kind) seems to be a natural consequence of basic intuitions underlying the "naive" set theory and referring to what can be called "Cantor's absolute".

The reader may think that the "natural" intuition mentioned was shared by Tarski; after all, as we have seen above, he did not renounce to work with any entity involved in classical mathematics, in the Platonistic sense of "entity". But things are very different, as this is followed by:

> On the contrary, we see at this moment no cogent intuitive reasons which could induce us to believe in the existence of cardinals $> \omega$ that are not strongly in compact [weakly compact], or which at least would make it very plausible that the hypothesis stating the existence of such cardinals is consistent with familiar axiom system of set theory.

It seems then clear that the "naive" intuition mentioned was not his intuition, and that some of the consequences of the belief in "Cantor's absolute" seemed to him, not only undecidable by the usual axioms of ST, but even probably inconsistent with them. As we have seen above, and we shall see still in a clearer way below, Tarski was a hardcore empiricist, so it is very likely that his intuitions about mathematical objects were mostly inspired in empirical objects, in those objects that can be visualized or imagined according to the model of the physical objects, and of the actual sets of them. That is why we called his type of mathematical intuition physical intuition.

Therefore, "naive" mathematical intuition must have seemed to him something not reliable, as it may lead us to entities which are not duly justified. Hence the tone of the reference to naive intuition and Cantor's absolute, which may well seem even of an ironic character. If this interpretation is correct, for Tarski classical mathematics should not embrace large cardinals, or at least some of them, the very large ones, and "believing" in a mathematical object should involve some connection with the empirical world.

This seem to me plausible according to the way he finished the paper: we need just add to the usual axioms of ST the statement that every cardinal $> \omega$ is strongly incompact [not weakly compact] "to preclude the existence of 'very large cardinals'" (*ibid.*). He does not openly defend such an option, and even if "very large" cardinals were eliminated, there could still be other large cardinals, but the way in which he describes those who may be opposed to the new, limitative axiom is very suggestive:

> Such a decision, however, would be contrary to what is regarded by many as one of the main aims of research in the foundations of set theory, namely, the axiomatization of increasingly large segments of "Cantor's absolute".

Thus, it seems clear to me that he did not consider himself among those "many", even more so according to the way he closes the paper with an implicit criticism of those who are ready to accept new "construction principles" and existential axioms, while "are not prepared to accept any axioms precluding the existence of such cardinals" (*ibid.*). So, the implicit reference to Gödel's Platonism seems once again clear enough. As a result, this may show Tarski's inclination not to be among those who favor stronger axioms for ST, but rather among those who eventually might favor limitative axioms instead. We will return to this issue later.[22]

If my interpretation is correct so far, the 1962 paper would be Tarski's first attempt to supply strictly mathematical results as grounds to defend a philosophical position, even though this is made in perhaps a very subtle way (the conviction and caution attitude again!).[23] And the most curious thing is that Tarski only returned to technical work in ST, in the 60's, just to try to state results, some of which may be seen as dealing with what we can call the ontological credibility of certain large cardinals, and of certain existential set theoretical axioms to justify them. So, the philosophical implications of these results seem to me highly relevant to understand the way in which he might have been concerned about the relation between the technical work on the least intuitive part of ST, the highest infinite, and his intuitive notion of set.

Gödel had suggested in his 1947 paper on CH that the very concept of set may require that the axioms of ST are extended by axioms of infinity to cover very large cardinals, and then (Gödel 1986, p. 182):

> Very little is known about this section of set theory; but at any rate these axioms show clearly, not only that the axiomatic system of set theory as known today is incomplete, but also that it can be supplemented without arbitrariness by new axioms which are only the natural continuation of the series of those set up so far.

So, Tarski's 1962 "philosophical" remarks were not passed unnoticed by Gödel, who made some comments about it in a footnote to his 1940 famous memoir about CH, added in 1965 (Gödel 1986, p. 97, fn. 2). There Gödel describes what Tarski did by saying that some new, strong axioms of infinity were there "formulated and investigated", then he seems to agree with Tarski in that those axioms are not supported by set theoretical intuitions. Yet he adds that even so the truth of those axioms is

---

[22] Tarski's usual tension between nominalism and realism appears here once more time: while he is somehow attacking the ontological exuberance of large cardinals, he does so by using non-finitary reasoning, based on infinitary languages, despite these languages should be rejected by his finitary "instinct".

[23] Kanamori (1997, p. 42), in noticing the philosophical flavor of Tarski's language here, quotes him at length, then describes Tarski's words as "guarded", and mentions the possible, realist influence of Cantor and Gödel. As we are seeing in the main text, things are much more convoluted from the philosophical viewpoint.

supported by "strong arguments from analogy", and he mentions certain technical result in the field.

Nevertheless, I think the agreement must have been superficial, for as we have seen the end of the Tarski paper clearly suggests that he might favor even limitative axioms instead of existential axioms for very large cardinals. Probably that is why when Gödel quotes Tarski's words "basic intuitions underlying the 'naive' set theory" (see the quotation above) he does it incorrectly, as he actually writes: "basic intuitions underlying abstract set theory"! In this way he replaces "naive", a word which he probably disliked as applied to his own form of mathematical intuition, by "abstract", a much more neutral, so acceptable a term.

Given Gödel's perfectionism, this must be due, not to a lack of care in quoting, but rather to some sort of *lapsus calami*, probably due to his unconscious Platonistic tendencies. It seems to me quite clear that Gödel noticed Tarski's implicit attack against Platonism, as well as his calling "naive" to certain intuitions about ST which Tarski did not share. Thus, although Gödel and Tarski may have shared that no form of reasonable intuition may lead us to so large cardinals, it is clear to me that for Gödel mathematical intuition was much more powerful and reasonable than for Tarski, who saw set theoretical "intuition" as rather unreliable a tool. So, I think we can accept as a result that Tarski distinguished between naive intuition, which for him must have meant arbitrary, artificial, Platonistic intuition, and "cogent" intuition, which I take to mean solid, reliable intuition, somehow related to the physical world.[24]

However, if we had available just the above evidence, our understanding of Tarski's intuitive notion of set would still be rather incomplete. Fortunately, there are two more pieces of unpublished evidence, which can be found in the Tarski papers: Tarski's contributions to two meetings which took place in 1965, in Chicago and London. Both contributions add important insights to his formerly known opinions on ST. In the first case, there was no proceedings published, but Tarski did not do anything to publish the transcription of his contribution. In the second case, it seems he did not allow the organizers to publish his contribution.

12.6 The Contribution to the Chicago Meeting

Tarski's contribution is the transcription of a tape containing what Tarski said in Chicago in 1965, at a joint meeting of the ASL and the APS (April 28–9, Marcus chairman), and in particular in a Symposium about the philosophical implications of Gödel's incompleteness theorems (Tarski chairman). The invited speakers to that symposium were Montague, Putnam and Benacerraf. Montague gave a brilliant, yet difficult talk about some sort of generalization of Gödel's incompleteness results;

[24] In 1970 Tarski mentioned to Mycielski "the Platonic belief of Gödel that sets can be seen (seen, not imagined) in our minds almost like physical objects", and that this belief "is bewildering" (Mycielski 2003, p. 217). It is difficult to find a most clear rejection of classic mathematical intuition, so I believe this should be taken as a further confirmation that for Tarski naive intuition was not reliable at all.

Benacerraf presented a precursor of his "God, the Devil and Gödel", published in 1967, then Putnam probably read a precursor of his celebrated paper "Mathematics without foundations", published in 1967.

In the transcription, Tarski is referring to those three talks, but he only mentions Putnam's name, and devotes most of his contribution to comments on Putnam's talk, although he obviously takes advantage to expose his general philosophical position, probably for the first time in public. He says to be an extreme anti-Platonist, a nominalist, a materialist, and clearly rejecting all a priori knowledge, from his empiricist and physicalist tendencies. That must have sounded as a bomb to many in the audience, who did not know anything about Tarski's actual philosophical tendencies, which were clearly strange given Tarski's very well-known works in ST, metamathematics, semantics, algebra, and many other parts of "high" mathematics. Since I am planning to study Tarski's general, philosophical ideas in a further paper, and since this contribution, together with the one to the London meeting (see below), will eventually be published, I will limit myself here to analyze the materials dealing directly with ST.[25]

In the text there are just two passages where Tarski seems to be reacting to Cohen's 1963 results, which were very recent at that time. In the first one, and after mentioning Putnam and the possible discussion between a Platonist and an anti-Platonist, he goes on to say that the anti-Platonist referred to in Putnam's talk "was of a very mild genre", and that he is "a much more extreme anti-Platonist". Then we find this:

> So, you see, I am much more extreme; I would not accept the challenge of Platonism. You agree that continuum hypothesis has good sense, it is understandable. No, I would say, it's not understandable to me at all. I would not be able to use like your anti-Platonist the terms "can", "be able", in such a way that it has no time coefficient. Now, one could think that from this point of view the whole problem vanishes. If one doesn't understand, then why bother when certain strange phenomena appear, strange things happen. Well, it doesn't vanish entirely; it takes another aspect. You think why other people are so impressed by it, why other people use it, why is it so that this Platonic way of thinking found such a firm ground in our society, in our culture, seems to have been very useful and very helpful in the development, in the progress achieved.

First of all, when he says that CH is not *understandable* to him, I think we can interpret him as meaning that his intuitions about the notion of set, his physical intuition, did not support a hypothesis so far away from what we know and believe about the physical objects, the actual sets of objects. That would be consistent with the former uses of "intuition" that we saw above when dealing with the large cardinals

[25] Thanks to Paul Benacerraf I was able to somehow reconstruct the information about the talks actually given in that meeting. Unfortunately, Hilary Putnam told me he does not remember anything about that forty years ago event. It seems that nobody replied to Tarski's shocking remarks, which is not strange; as Benacerraf said to me in correspondence, although he was taken aback by what he heard, and of course he did not agree with it, Tarski was among the gods, and no one replies to a god! Tarski's philosophical position was so unexpected that Benacerraf even "forgot" it when he wrote "Mathematical truth", published in 1973, where Tarski's position is taken to be quite the opposite to the one made public in the meeting. [The full text of this contribution was published by this writer here: "Two unpublished contributions by Alfred Tarski." *History and Philosophy of Logic* 28 (2007), pp. 257–264.]

issue: intuition must have meant to him something similar to visualization, to our capability to imagine something in a similar way to which we imagine something physical. That is why we called it physical intuition. Therefore, Tarski seems to me to be thinking of his well-known ultimate indistinction between logical and factual truths, between mathematics and physics.

But what about the "time coefficient" puzzling expression? When I read parts of Tarski's contribution in a seminar in Berkeley,[26] Henkin told me that the reference to a time coefficient was not to be understood in the intuitionistic sense, but just referring to something which to discover a yes or no response is a matter of time. This could mean that for Tarski talk about CH in general, abstract sense, had no meaning at all unless it is inserted in a particular historical time, then related to the state of the research on CH in that moment. And this will more or less coincide with my former interpretation about understandability of an abstract, mathematical hypothesis rather in a proof-theoretic sense (see above). So, the "yes or no" issue by Henkin is not to mean anything about "constructibility", but just about the insertion of a hypothesis in a more general, scientific context, which may well be historically changed.

To my knowledge, the expression "time coefficient" appears just one more time in Tarski's writings: in his 1944 divulgative paper on the semantic definition of truth. There he says that acceptability of an empirical theory or hypothesis must be relativized "to a given stage of the development of a science": "In other words, we may consider it as provided with a time coefficient: for a theory which is acceptable today may become untenable tomorrow as a result of new scientific discoveries" (Tarski 1944, §21).[27]

So, the last part of the quotation seems now to be clearer. Even when we declare CH to be relative to a more general body of global knowledge, the problem does not vanish "entirely" but it takes another aspect, i.e. it goes far beyond a question of pure mathematical provability. Once we insert the problem of the understandability of CH in a historical context, where mathematics is closely related to other scientific theories, and then is treated as an empirical science, in the sense that its hypotheses have to be evaluated in a more general context than pure mathematical provability, as for instance fruitfulness of consequences, and useful relations to our empirical knowledge of the world, then the Platonistic tendency shows its only hope of an actual justification: pragmatic justification. So, Platonism, if accepted in some ways,

[26] In June 1996, Mathematics Department, with John Addison, Charles Chihara and Leon Henkin. That session was not only very insighful for me, but also a big pleasure, as it took place in the same room, and chairs, where Tarski used to meet his colleagues and collaborators when he was in his department.

[27] Also, Frost-Arnold suggests in correspondence the following, just as a conjecture: "The 'time coefficient' stuff may be related to the 1940-41 conversations. For there, Tarski says a finitist-nominalist must get rid of 'x is derivable/ provable', because it is an irreducibly modal notion, and replace it with 'x is derived/proved'. But at any given time, only a finite number of theorems will ever be proved/ derived, and what is proved will always be relative to a given time". This is obviously food for thought, yet I still like my own interpretation best, mostly because it is based on textual evidence.

should be considered as a heuristic tool, because it helps us to expand our view of the world, as long as we regard it as just a working tool, not a body of beliefs in a transcendent world. Thus, the entities which we arrive at by using that tool should be calibrated with our usual, pragmatic, relative standards, always closely connected to the empirical, the physical only actual world.[28] This seems to me to be compatible with Corcoran's proposal of a Tarskian "Platonistic reductionism".

As a matter of fact, Tarski himself admitted, in conversation with Chihara, that he accepted that for Gödel his results were important for philosophical reasons, and that a nominalist could hardly have discovered them (Chihara 1996, personal communication). So, it is clear that Tarski accepted that Platonism could be used as a heuristic tool to reach certain results which are *prima facie* very far beyond our physical intuition. Yet this could also be regarded as a rather puzzling assertion: did Tarski mean that he could not have reached Gödel's results by himself, in being a nominalist? Hardly. First because he was actually convinced that he was very close to have reached those results himself, and that with more time for research this could probably have been done. And, most interestingly, because as a working mathematician he was a methodological Platonist, no matter he may have thought that ultimately his abstract entities could somehow be reduced to physical ones.[29]

The second passage can be found at the very end of the contribution, i.e. after Tarski already gave his explanation about the probable historical origin of logical truths (so expanding what he wrote in the 1944 letter to White, published in Tarski (1987)), and after his assertion that the laws of logic (especially the law of excluded middle), which were born to deal with physical objects, were then wrongly applied to abstract notions. Then he adds:

> It seems now pretty certain [than? then?] there are expressions in set theory which we consider sentences —this means expression of truth or false— and regarding which we shall never have (it seems likely at least) any possibility of deciding whether they are true or false. And the natural question arises whether it still makes sense to stick at [to?] this logic, old logic, when dealing with this type of sentences. So in this sense I would not go as far as you —than in mathematics really truth is only provability. I don't know which provability, but at any rate maybe the notion of truth is simply not proper, this classical notion of truth, for mathematical sentences. I think that this is the negative conclusion which one could draw from these developments starting with Gödel, and therefore I am quite interested in attempts of [at?] constructing set theory on the basis of some non-classical logics, simply as an experiment. We shall see to what it will lead.

First, he mentions sentences which we could call absolutely undecidable, by following the terminology he used in his letter to Gödel of 1946 (see above). In the context of the contribution this was probably done by thinking of CH, especially by taking into consideration the former quotation and Cohen's famous results from 1963 about the independence of CH from ZFC.

---

[28] Including the mental, since for Tarski, following Kotarbinski, the psychical should be somehow reduced to the physical.

[29] For details about Gödel's view of Platonism as an heuristic tool, see my Rodríguez-Consuegra (1995). Also, it contains a comparison between Tarski, Gödel, and others, concerning the similarities between mathematics and physics.

Second, since the quotation immediately follows his claim that the law of excluded middle was born in physical, historical contexts and was then wrongly applied to abstract notions, his question about applicability of the old logic may be referred to classical, bivalent logic, the logic with just two truth values. So, when he finishes by asking for a formalization of ST in a non-classical logic he might have been thinking of some system of three valued logic, as the ones used by some of his teachers in Poland (mostly Lukasiewicz). And this is interesting, for it allows us to relate the issue again to his well-known attack to the factual-logical distinction, as applied to the truth of sentences in his 1940–1 conversations with Carnap, and also to his well-known reference, in his 1944 letter to White, that logical laws are not different in nature from empirical truths, but only in degree. They are just more general and also historically older, so some empirical, fundamental experiences may make us to be inclined to change them (and he mentions quantum mechanics).[30]

Finally, the issue that truth is not just provability. As this is apparently referred to Putnam's talk, I looked for any such assertion in Putnam's actual publication of "Mathematics without foundations", and it is not to be found there, though the paper as published might be different from the paper actually read by Putnam, and even the transcription of the tape might not be totally fair here to what Tarski actually said.

Anyway, Tarski was probably inserting Putnam's general talk in the classical context that in mathematics the only criterion to be used for the working mathematician is provability, including of course Tarski himself (pace Gödel's incompleteness and undecidability results). But even if it was totally clear what provability should be (which seems not to be clear for Tarski as per Putnam's actual talk, were modal notions are used, which Tarski was not very fond of), then provability is not enough, in the sense that the classical notion of truth, the bivalent truth, is supposed to underlay our results when applying the standard provability methods. And as we have seen for Tarski the law of excluded middle is rather suspicious when applied to hypothesis like CH, which are very far beyond our physical intuition. Thus, when he writes that his negative conclusion about classical truth started with Gödel's results he must have been thinking, not only in the obvious impossibility of continuing to identify truth with provability after them, but mainly in the similarity between logico-mathematical truth and factual truth that was so dear to him. Therefore, the analysis of the second quotation lead us to very similar conclusions that the ones reached with the first quotation.

[30] It is also possible that Tarski was thinking of some form of intuitionistic logic, perhaps in order to dispense with the law of excluded middle, as in his writings there are a few places where he pointed out some intuitionistic flavor of his work, but for him intuitionism limited classical mathematics in an intolerable way, so this possibility seems rather unlikely.

12.7 The London Meeting

About three months after the Chicago meeting, Tarski participated in the 1965 International Colloquium in the Philosophy of Science, held at Bedford College, London, from July 11 to 17. One of the main sections of that meeting was devoted to Problems in the Philosophy of Mathematics, whose proceedings were published in 1967, edited by Lakatos (1967). Due to the spectacular results by Cohen in 1963, two sessions were devoted to the recent results in ST, with talks by Mostowski and Bernays, followed by rather hot discussions, especially after the first one, where Tarski was the chairman. Also, Kreisel gave his famous talk "Informal rigour and completeness proofs", and heavily contributed to the discussions, especially attacking Mostowski's presentations and ideas.[31]

The gist of the context of the discussions, in connection with Tarski's contribution, seems to me to be this. Kreisel defended the role of second order logic in formalization, and pointed out that there is an asymmetry between the independence of CH and that of the parallel postulate in geometry, for the latter is also independent in second order geometry and the first is not: CH is decided in second order ZFC.[32] Also, he attacked Mostowski's saying that the notion of set is vague, and the "old-fashioned" belief, either realist or idealist, that what the mathematician does is analyzing intuitive notions then establishing their properties. On the contrary, he defended "informal rigour" and the need to introduce new, stronger axioms to decide important problems, by using second order logic when needed.[33]

Second order logic was criticized by Kalmar and Mostowski. The latter admitted that CH is decided in second order ZFC but added that this is useless as no one knows how it is actually decided. Also, he said that second order logic is a part of ST, so trying to develop both at the same time makes no much sense. Finally, Mostowski insisted that the present situation of ST is problematic, and that it will probably lead to a bifurcation into different set theories (by which he must have been referring, for instance, to Cantorian and non-Cantorian set theories, i.e. theories in which CH is added to the usual axioms and theories in which it is not, or to similar bifurcations as a way to absorb the recent, controversial results in ST).

Tarski made a contribution, probably after Mostowski's talk, in which he explicitly mentions the discussion between Kreisel and Mostowski. His contribution was not

---

[31] Mostowski was a former student of Tarski in Warsow, under whom he wrote his 1939 Ph. D. dissertation, although the official supervisor was Kuratowski.

[32] Kreisel apparently did not know that Zermelo pointed this fact out already in 1930; see Moore in (1986, p 157), fn.

[33] Kreisel's global position shows a clear influence of Gödel on the need to introduce stronger axioms to try to decide CH and similar problems, as well as about the asymmetry between the independence of CH and that of the parallel postulate; see Gödel 1986, p. 267. In addition, Kreisel's contributions have a clear intuitionistic flavor, especially when he criticizes the belief in a previous, intuitive notion of set, independent of any formal description of it. In this connection Kreisel's position seems to me very close to Dummet's well-known arguments that there is no model of mathematical entities, e.g. of the notion of set, apart from the precise, theoretical description of it; see Dummett (1967).

published in the proceedings, but a transcription, probably from a recorded tape, is in the Tarski papers in Bancroft Library. I suppose he did not allow the organizers to publish his contribution, but the reason is unknown, although he was probably not satisfied with it, perhaps because it showed too much of his true philosophical tendencies (if so, this may well be a further sign of a "conviction and caution" attitude). The following is an attempt to analyze its content, but just in what is relevant to his notion of set.[34] This may well be the darkest scenery about ST described by Tarski, with the possible exception of the Bucharest talk in 1971 (see below). In the contribution, Tarski discussed two main questions: second order formalization and the present state of ST, to which the following two sections are respectively devoted.

12.8 Second Order Formalization

Tarski referred to it as "the use of the notion of set in the second order formalization of an arbitrary theory".[35] He starts by pointing out that first order formalization is very useful to formalize a large part of mathematics, and that in doing so it "has gained a very wide range of agreement, but not an absolute one" (p. 2). Then he mentions certain objections as having been formulated (presumably by Kreisel in his well-known paper read there), and he illustrates the objections with the example of first order formalization of natural numbers theory in the usual way: the axioms, followed by formalizations of proof, theorem, etc. Then the objection may be raised that in formalizing the notion of proof "you will again use notions such as 'set' whose exact meaning is by no means clear to you" (p. 2), which is not an elementary notion. Then he said:

> Now I would answer that this use of non-elementary notions is not necessary in formalization. I have often stressed that it is not necessary in formalizing (if by that one means normal first order formalization) to define what is a formula, what is a proof, what is a theorem. Formalization may consist in a series of autobiographical statements. Of course the statements must include such sentences as: "I recognize the following statement as true, as an axiom, and I don't want to analyze it" (and I write down the axiom). Then I say, "I shall also recognize as true any sentence b under the condition that I have previously recognized as true two sentences, one of which is a and the second of which is $a \supset b$." And this is enough. You see that in this formalization I do not actually use any such theoretical term as set; and I do not need a definition of a sentence or a formula. I have said quite clearly what I am going to do in developing arithmetic, and this procedure is quite satisfactory for various purposes.

Several things are remarkable here. First, the notion of set is admitted being a non-elementary notion, so requiring a theory, ST, to give it a precise meaning, and perhaps

[34] Solomon Feferman and Ignacio Jané helped me dating the transcription. Jané finally suggested the London 1965 meeting, then everything was on the right place. This contribution will eventually be published, together with the contribution to the Chicago meeting. [The full text of this contribution was published here by this writer: "Two unpublished contributions by Alfred Tarski." *History and Philosophy of Logic* 28 (2007), pp. 257–264.]

[35] This is interesting because Tarski is using the word "notion", presumably as opposed to "concept", and it is well-known that he considered "notion" to be hopelessly vague.

to transform the notion into a clear concept. Second, the mixture of formalization and axiomatization seems to me rather unclear. If you just offer a list of autobiographical statements which you postulate as true, this is just an axiomatization, though of course you can also formalize those statements later, for instance by using first order logic.

But of course, he goes on, the use of "theoretical terms" from ST is unavoidable in formalizing other theories, as can be seen by trying to state the simplest example: the rule of induction. And the same could happen with a second order theory:

> Now this is so in this case, and if in this case you still admit that the formalization by these means of the rule of induction clarifies to some extent a certain theory, then I think we should also recognize the fact that *other* set theoretical devices may be useful in formalization: and second order logic is just such a device.

In view of this passage, Tarski's acceptation of second order logic seems clear. However, his use of the logic of PM in many of his works, which involves higher order logic in that it uses variables of many different types (not just individual variables), was already well-known. So, the question is: why is Tarski saying so? It seems to me that in the context of the discussion between Kreisel and Mostowski he was inclined to share Mostowski's rather suspicious attitude before second order logic, as what he says is that second order logic must be regarded as a set theoretical device, and he even emphasized the word "other", to make clear the power of ST as a supplier of theoretical devices. But this is exactly the position showed by Mostowski when he said that second order logic is but a part of ST (see above). Tarski was very polite in the introduction of his contribution: he said to be "close" to Kreisel, but made very clear that he knew that there was no disagreement between Mostowski and himself. What was turning out instead was rather that he did not like second order logic as an instrument on its own.[36]

12.9 The 1965 State of Set Theory

As for the second question, Tarski seemed to be very interested in pointing out that the unsatisfactory state of ST may be due to the notion of set itself, and to the way we "think" we understand that notion. First, he says that in that discussion "we all use terribly loose terms, all of us, and my discussion will not be on any better level, or on any different level, than the discussion of others" (p. 1). Then he goes on to develop this impression like this:

> We have heard a lot about the unsatisfactory situation at present in set theory, about the possibility or ramifying set theory, and that in fifty or a hundred years there will be no single set theory, but many set theories, so that set theories will be like groups, for example. Maybe

[36] A further sign that Tarski "came to doubt the foundational significance of higher-order logic" is that his famous 1959 paper entitled "What is elementary geometry" was devoted to the construction of a first order foundation for geometry, so "correcting" former second-order attempts in the 30's (Corcoran 1991).

> this will all come true, but it will not mean that set theory as such will vanish, because in developing set theory we certainly have in mind some aesthetic model, or rather some rule which singles out a privileged class of models, since we do not know exactly what model we have in mind.

For one thing, it may be shocking for many to read that for Tarski the possibility of different set theories, depending of the use of special axioms or hypothesis (like CH for example), may be regarded as unsatisfactory. After all, his pragmatic position in philosophy was close to formalism (and no longer to Lesniewski's "intuitionistic formalism"), and his general philosophical position was a clear nominalism. So maybe he was just showing the optimistic side of his Platonistic reductionism here.

Also, his skepticism about the vanishing of ST in the future, due to the consequences of Cohen's and similar results, is not consistent with what he said to Carnap in the 40's (see above), that it would be a guess and perhaps a wish that "general" ST will disappear in the future. In 1965 part of the future had already arrived and we find Tarski defending classical ST with classical arguments: that we have "in mind some aesthetic model" of set!

However, how do we know that we have a model of set in mind? In the context of Tarski's philosophical ideas, this must have been referred to our empirical knowledge of sets, to our physical intuition, as he rejected any a priori knowledge or any mathematical intuition in the Gödelian sense. Thus, to qualify that model as "aesthetic" should be understood in the context of our perception of physical objects, shapes and forms: as something we can somehow visualize or imagine, as we imagine physical objects not present before us.[37]

If this is right, some of the reasons for not allowing the contribution to be published might be emerging here, as Kreisel was precisely strongly criticizing the "old fashioned" conception that we first discover a model in our mind then we make it precise with a mathematical theory, e.g. through a set of axioms. We do not know whether Kreisel did reply to Tarski in that discussion, probably not, but it would have been very easy for him to use the Dummett (1967) style arguments against Tarski's old-fashioned position: we do not have any access to those models independently of our theories, because we cannot make them precise without those theories. After all, Tarski may well be recognizing this, at least in part, when he closes the passage by speaking, not of a particular model, but of a whole class of them, "since we don't know exactly what model we have in mind". But this is still worst, as the only way to characterize a class of models is a set of axioms, and axioms are the raw material of deductive theories.

All in all, it seems to me that Tarski was trying to be too cautious in connection with what he probably believed: that we possess an intuitive notion of set which develops in contact with the physical sets of objects laying around us, in our ordinary relation and interaction with the world, and this notion is very different from many other mathematical notions. In this sense physical sets are finite, so a finitistic

[37] The strange use of "aesthetic" here could perhaps be also related to Tarski's love for classical mathematics, or to his attraction for the beauty of mathematical theories, as ST; see above in the conversations with Carnap section. Yet I still prefer the interpretation in terms of visualization.

position was always sympathetic to Tarski, as can be clearly seen in many passages of his conversations with Carnap, as the one quoted above. Otherwise, if we interpret his words as supported by the implicit acceptation of some form of mathematical intuition, in the Gödel style, then Tarski's position would not be very different of that of a literal Platonist, so not the words expected from a Platonistic reductionist. Everyone who worked close to Tarski knew very well that Tarski rejected "classical" mathematical intuition. As Givant wrote to me:

> Gödel was in some sense a Platonist. He thought deeply about the Axiom of Choice and the Continuum Hypothesis, and believed the former to be true and the latter to be false. Tarski did not share these "intuitions". For him it was not evident that the Axiom of Choice was true, any more than it was evident that the Continuum Hypothesis was false, and he did not comprehend Gödel's intuitions. At one point I heard Tarski express some sentiments that might even be interpreted as ultrafinitistic.

Let us continue with Tarski's text. The differential character of the notion of set, as compared with many other mathematical notions, or concepts, is also that we use it in ordinary language, in ordinary life. That is why Tarski insisted that we should not regard ST as any other mathematical theory, because "the fact that we permit ourselves to use ST in non-formalized ways in everyday language shows, I think, that ST is very different from group theory" (p. 4). And the same point was made before (p. 3):

> The fact that we use set theoretical notions in metamathematical or metalogical discussions seems to me very important, because in such discussions we use the word 'set' in a non-formalized way. Thus there is a difference between set theory and other kinds of mathematical theories.

This is perfectly consistent with my above interpretation in terms of physical perception: we do "perceive" sets, but we do not perceive groups, or vectorial spaces, and the like. The historical basis of the most elementary truths of logic, in close contact to everyday life, must have been for Tarski very similar to the case of the historical origins of ST: our empirical relationship with physical reality. That is why we understand the notion of set in a previous manner, without any recourse to mathematical theories. So, the way in which we possess a model of set in mind is just a visual way, a spatially imagined way, similar to a physical object, no matter how complicated or convoluted. And this seems to me the only alternative open to us if we do not want to make of Tarski a true Platonist, not just a methodological, a tactical one. (Yet see the last section below for more problems about perceived sets.)

Also, it allows us to better understand the impact of Cohen's, and similar results: in making the "bifurcation" of ST possible, Cohen's results showed that our knowledge of the intuitive notion of set, as reflected in classical ST, is not precise. If we have to allow the ramification of ST, and the appearance of many different set theories, what convincing basis would have we to say that our physical intuition is reflected in this or in that particular ST? Here Tarski might be facing a similar problem that the one which led him to consider higher logics as ultimately artificial, or to say that there is no difference in nature between mathematical theories and physical ones, or to defend that in applying classical logic to abstract entities we are forgetting the historical, empirical origins of logical truths. But if classical ST can be defended just by appealing to physical intuition, the notion of set can no longer be respectable to mathematicians! And this is, I think, the big problem for Tarski, who always wanted to make logic and metamathematics acceptable to mathematicians as full mathematical theories. If bifurcation of ST is confirmed, then the notion of set is hopeless. This leads us to the end of Tarski's contribution, which is really gloomy.

The problem is, according to Tarski, that when ST is used in metamathematical discussion "we rarely ask anybody to explain the set theoretical notions". And this is a problem, because very different set theoretical notions can be used, and some of them may lead us to "deception". Tarski gives an example: someone might prove the categoricity of a theory by using "a restricted naive set theory" with no inaccessible cardinals, so his proof, in starting from an "artificial assumption" is a kind of deception. Then Tarski finished the contribution this way (pp. 4–5):

> For this reason I consider the situation in set theory to be even more unsatisfactory than do other people. I am not consoled when I am told that the time will come when there will be many set theories, many different models, and so on. Recent results in set theory show that there are many set theoretical models for our usual set theoretical axioms. And the problem of whether or not there is one optimal model remains open; it is a problem of fundamental importance.

To be noticed first is Tarski's reference to inaccessible cardinals, whose content does not seem to coincide with what he wrote in the Stanford 1960 paper (Tarski et al. 1962). As we saw above, in that paper Tarski seemed to somehow favor a limitation of the more exuberant extensions of ST to very large cardinals, in particular to some inaccessible cardinals. Also, he seemed to call "naive" ST one which is not restricted in any way, so perfectly able to pursue "Cantor's absolute". But now he makes clear that such a limitative restriction to eliminate inaccessible cardinals would be an artificial assumption, for it may lead us to unreliable proofs. And worse, he now calls "naive" precisely to the ST resulting from such limitations, while under our former interpretation above naive ST was the totally unrestricted one, which was to be somehow rejected from the viewpoint of Tarski's physical sense of intuition. So, we are in trouble.

A first attempt of a solution might go like this. Our sane intuitive notion of set, the one based on physical intuition, is the only one capable to provide us with an ontologically reliable basis for the ultimate reduction of the whole of higher mathematics to something acceptable from a finitistic viewpoint, or at least to an infinite basis which seemed to him somehow reducible to the former. But this physical intuition does not allow us to go very far, as it is essentially limited to finite languages and Tarski had no intention whatsoever to renounce to the beauty of classical mathematics. So, if we limit the usual ZFC too much, that would force us to renounce to be able to manage with important theories and proofs and with their metamathematical properties. Thus, while his Platonistic reductionism pushed him to restrict the ontological extension of ST, his needs as a working mathematician pushed him in the other direction. And this dilemma has no solution, as the motto of this paper shows.

But what about "naive" ST? Is this the unrestricted or the restricted one? Well, Tarski might be calling naive to any ST whose axioms are somehow determined by a priori beliefs. Belief in the full Cantor paradise is an a priori belief, but any restriction of the usual extensions of ZFC to very large cardinals, without any basis in physical intuition, could be a priori as well, even when it might favor the strong limitations dictated by our intuition of physical sets, in the sense that any restriction whose resulting entities are not based on, and justified by, a well-known process of reduction might be arbitrary then artificial. Unless we accept something like this, the only other alternative would involve limiting the range of the Cantor absolute, and this, again, would be contrary to the distinction between naive intuition and physical intuition.

Of course, some inconsistency on Tarski's language is always possible, especially because he was no professional philosopher nor was he very concerned about developing his philosophical options in much detail. But he was obsessed about the use of a very precise terminology, so I think he was aware of his former uses of the word. In addition, I tend to believe that his philosophical inclinations, though not very sophisticated, were tremendously permanent and invariant (even under every permutation of the world onto itself !).

Be that as it may, the end of the contribution is totally clear: the reason of Tarski's judgment about the unsatisfactory situation in ST was not only based on Cohen's results, which are not even explicitly mentioned, and on the ramifications which might result from them. (As a matter of fact, he was convinced about the absolute indecidability of CH in ZFC since at least 1946; see above.). But also on the problem of the legitimate extensions of ZFC: in the question whether or not we are to allow the addition of strong axioms, or of limitative ones, to the usual axioms of ST. So, when he talked about the problem of finding the "optimal" model as a fundamental problem he was totally serious. The optimal model would be the one which, while providing us with a ST capable to be consistent with our physical intuitions of sets, it would nevertheless be clear enough to decide between too large and too short extensions of ST without resorting to a priori, or theological, options. Therefore, the main problem remains open: where to look for that optimal model?

When I sent the text of the London contribution to Solomon Feferman he wrote to me: "These remarks indicate the extent to which Tarski accepted set theory and 'core' set theory (e.g. ZFC) as expressing a definite concept". I fully agree, but to what extent those remarks show Tarski believing in such a definite concept? I think he believed that ST, as we know it, is hopeless to grasp such a definite concept. But this was no problem for him, as he did not believe in the existence of concepts at all! He was a nominalist, so he believed that in contact with our physical intuition of sets we develop just set theoretical notions, but that unfortunately those notions are not enough to develop higher mathematics without using Platonistic entities. And this tension has, again, no solution. Through the independence results, and the problem whether or not ZFC should be extended with no restriction, it seems to me that for Tarski the standard axioms of ST show some sort of essential indeterminacy, or relativity (to use Skolem's word): they are not the sort of thing capable to transform our intuition of sets into something precise, let alone into a Platonic concept, so our intuitive notion of set is to be kept essentially indeterminate.

12.10 Are Set Theoretical Notions Logical?

In his semantical and mathematical practice of the 30's Tarski applied a rather broad sense of logicality, where many notions not *prima facie* belonging to logic were described as exhibiting a "general logical character" (Tarski 1983, p. 170). Among them he clearly included set theoretical notions, particularly membership, and even some from the domain of cardinal arithmetic (p. 171). This was justified this way: "The calculus of classes is a fragment of mathematical logic and can be regarded as one of the interpretations of a formal science which is commonly called Boolean algebra or the algebra of logic" (p. 168). Yet as we have seen in former sections of this paper, this procedure involved the standard use of a simplified type theory, and although type theory offered mathematical universality, it made true domain universality impossible, so type theory was finally abandoned in favor of standard ST, at least for the definitions of the most fundamental semantical notions. Thus, in dispensing with type theory, the general logical character of set theoretical notions was no longer clear.

In the famous 1944 letter to White, an ambiguous attitude about the issue is apparent: "sometimes it seems to me convenient to include mathematical terms, like the ∈ -relation, in the class of logical ones, and sometimes I prefer to restrict myself to terms of 'elementary logic'. Is any problem involved here?" (Tarski 1987, p. 29). But it was not until 1966 that he returned to the issue in a more detailed way.

In that year Tarski gave a lecture in London, entitled "What are logical notions?", which was later edited by John Corcoran and appeared published in 1986 (Tarski 1986b). The text is now well-known and have been deeply discussed, but it contains something very relevant to Tarski's viewpoint about set theoretical notions, so at least some brief discussion here is mandatory. I am talking about Tarski's question whether or not set theoretical notions, i.e. mostly the membership relation, are logical notions. The question is very important because the language of ST is mathematically universal, so capable of expressing the whole of mathematics, thus the real question is whether or not mathematical notions are logical.

According to Tarski's definition, logical notions are those invariant under all possible one–one transformations of the world (the universe of discourse) onto itself. Tarski's response to the question is ambiguous. If you use PM type theory, a clearly higher-order logical theory, membership is a logical notion, so ST, then mathematics, is a part of logic. The PM fundamental universe contains all individuals, "which we may interpret as the universe of physical objects" (Tarski 1986b, p. 152). Yet type theory requires that in addition to the most basic universe of discourse, that of individuals, you have to construct further domains containing classes, relations, and the like. On the contrary, if you dispense with the hierarchy of types and regard membership as an undefined, primitive relation between individuals, so that you only have a unique universe of discourse, then membership is not a logical notion (it is not a logical relation between individuals), so first order ST, then mathematics, is not a part of logic.

To finish, Tarski associates each method to a different type of mind: philosophers would favor the first one, characterized by a monistic conception of logic, ST and mathematics, while mathematicians would prefer the second, where mathematical notions and theories have an independent ontological status on their own. However, Tarski does not tell us which method he liked best at bottom ("As you wish"!), exactly as he already did in the letter to White, where he sees no problem in either possibility. So, we can speculate in view of our former discussions, just to see whether or not some option would be more coherent with which we have learnt from the other evidence above.

A further hint is contained in Tarski's contribution to a meeting in 1955, where he made some comments about Skolem's relativism about set theoretical notions (see Bellotti (2003, p. 406). The important point is contained here (I quote after Bellotti):

> The Löwenheim-Skolem theorem itself is not true for a certain particular interpretation of the symbols. In particular, if we interpret the symbol '∈' of a formalized theory of sets as a dyadic predicate analogue to any other predicate, then the Löwenheim-Skolem theorem can be applied. *But if instead we treat '∈' like the logical symbols (quantifiers, etc.), and we interpret it as meaning membership, we will not have, in general, a denumerable model.*

I take this to mean that when ST is constructed the second way, the axiomatic one, LS holds, but when we use, for instance, PM type theory, then membership becomes a logical notion, and LS does not hold. We should not forget that type theory uses a higher-order logic, and in second-order logic LS does not hold. Thus, if we regard ST as a part of logic it is released of the relativity problems intrinsic to axiomatic ST on its own. Bellotti takes Tarski's words to involve that he was finally decided to see membership on a par with the usual logical notions. I do not think this is right, since as we have seen in the first section above Tarski dispensed with type theory to complete his semantical program, but it could anyway be added to the things we need to consider trying to answer the question: which one of the two above methods should be chosen?

As we have seen in the first section above, if type theory is chosen, which is what the historical Tarski used to do in some of his writings, ST is a part of logic, but although we do have mathematical universality, we do not have domain universality (we do not have a single universe of discourse). Also, the fact that the basic domain, that of individuals, can be identified with the physical world, is a further advantage, as it fits very well Tarski's physical intuition. Finally, second order logic (and higher ones) is inevitably involved, so the former limitation of LS is just a partial one, as LS does not hold in higher order logics. And we must remember that Tarski did not like second order logic, and probably shared Mostowski's conception that second order logic is a part of ST! (see above).

On the other hand, if ST is not logical we still have mathematical universality, and also true domain universality, but individuals, as well as sets, belong to the same, single universe of discourse, since in first order ST sets can be individuals. Thus, physical intuition, acting mostly on physical objects, works just in an analogical way with sets of them, not in a literal way, as sets are only "individuals" in a metaphoric sense. Finally, the usual ST is a first order theory, and this is good, but then LS applies, with the well-known inconvenient consequences.[38]

Therefore, no option is satisfactory, as we do not get mathematical universality, domain universality, ontological credibility, clear intuitive access, and lack of indeterminacy at the same time, in any of them. I think that is why Tarski did not make clear which one was his own option: we may well be before another example of a tension, a philosophical tension, with no solution. After all, he was no longer very interested in ST and preferred, by far, a sort of algebra powerful enough to do the same job without any of the usual set theoretic problems, including indeterminacy.

In 1971 Tarski gave a talk in Bucharest, which had a rather philosophical content related to his conception of ST and its present state. In the talk Tarski insisted on the 1965 gloomy lines, but that time there was no tape recording what he said, and to my knowledge no transcription was made, so very little is known for sure. Givant did not attend the talk, but he spoke to some of the people who listened to Tarski there, then wrote this to me: "I believe people who heard Tarski's talk found it profoundly pessimistic. I would not say that his point of view was shared by many philosophers and logicians, though I would guess that it might have the tacit support of many mathematicians". So, his initial hopes that mathematicians accepted foundational endeavors as being fully mathematical seemed to end in a failure, at least to the extent that those endeavors were somehow based on ST.[39]

Acknowledgements I spent most of June 1996 in Berkeley, Mathematics Department, studying the Tarski archives in Bancroft Library, thanks to the financial support of my Spanish Research Project DGICYT PS93-220. During my stay there, I was so fortunate as to enjoy the personal help and advice of Leon Henkin, Charles Chihara, John Addison and Jan Tarski, as well as of Solomon Feferman and Patrick Suppes, in correspondence. After a number of years working mostly in other fields, I recently went back to Tarski's philosophical ideas, and the present paper is a first result.

---

[38] As it was mentioned above in the main text, in order to avoid the applicability of LS then of Skolem's relativism to the usual axioms of ST, Zermelo published a second order axiomatization of them in 1930, but obviously this could not offer the goods which Tarski wanted.

[39] This seems to somehow coincide with what James T. Smith, who actually attended that talk, has written to me:

> Detail on Tarski's gloomy Bucharest talk. I don't remember very much, but here goes. The most spectacular results since 1963 had been in metamathematics of set theory. That was generally the buzz in the corridors. (I do remember another buzz having to do with automata theory – Lindenmeyer systems, etc. – but that might have been just in the crew I was hanging out with.) Tarski spoke about the set theory results, and concluded that however spectacular and amazing they are, they may never lead to adequate answers about the foundations of mathematics. These are my words not his. I mean foundations in the sense of fundamental principles. I don't think he listed his questions, which would have revealed some of his philosophy. I think he was saying, you people won't be satisfied by these results, not that these results won't be answering his questions. I think he did not say anything like "these questions are meaningless" – that would have been less gloomy, for we'd have been freed of the problem. I think he said something to the effect that the questions – whatever they are – would have to be reformulated to admit satisfying answers.

Thus, if no results in ST would never lead us to the right answers to the questions involved in the foundational task, he might have felt that he wasted a lot of time and effort for nothing.

Many people have helped me, in correspondence, during the process of writing this paper, in different ways. Although I have mentioned some of them in the footnotes, let me mention all of them here as a way to express my gratitude: Paul Benacerraf, Charles Chihara, John Corcoran, Philippe de Rouilhan, Solomon Feferman, Greg Frost-Arnold, Steven Givant, Mario Gómez-Torrente, Ignacio Jané, Paolo Mancosu, Elena Marchisotto, Judith Ng, Hilary Putnam, and James T. Smith. Also, the kind permission of the Bancroft Library to quote from the Tarski unpublished materials is gratefully acknowledged. Finally, my thanks are also due to John Corcoran, Greg Frost-Arnold, Mario Gómez-Torrente, Ignacio Jané, Paolo Mancosu, and Roberto Torretti for comments over former versions of this paper.

References

Bays, T. (2001) On Tarski on Models. *The Journal of Symbolic Logic*, 66, 1701–1726.

Bellotti, L. (2003) Tarski on logical notions. *Synthese* 135, 401–413.

Bernays, P. (1967) What do some recent results in set theory suggest? In Lakatos, I. (ed.) (1967), 109–112.

Corcoran, J. (1983) Editor's Introduction to the Revised Edition. In Tarski, A. (1983), xv–xxvii.

Corcoran, J. (1991) Review of A. Tarski, Collected Papers, vols. 1–4. *Mathematical Reviews* 91, 91h:01101–04.

Corcoran, J. (1995) Tarski, Alfred. In Kim, J. and Sosa, E. (eds.), *A Companion to Metaphysics*. Blackwell, Oxford, 487–489.

de Rouilhan, P. (1998) Tarski et l'universalité de la logique. In Nef, F. and Vernant, D. (eds.), *Le Formalisme en Question. Le Tournant des Années 30. Vrin,* Paris, 85–102.

Dummett, M. (1967) Platonism. Reprinted in his *Truth and Other Enigmas*. Harvard University Press, 1978, 202–214.

Erdös, P. and Tarski, A. (1943) On families of mutually exclusive sets. *Annals of Mathematics* 44, 315–329. Reprinted in Tarski, A. (1986), vol. 2, 591–605.

Etchemendy, J. (1990) *The Concept of Logical Consequence*. Harvard U.P., Cambridge.

Feferman, S. (1996) Gödel's program for new axioms: Why, where, how and what? In Gödel '96 (P. Hajek, ed.), *Lecture Notes in Logic* 6, 3–22.

Feferman, S. (1999a) Tarski and Gödel: between the lines. In Wolenski, J. and Köhler, E. (eds.) (1999), 53–63.

Feferman, S. (1999b) Does mathematics need new axioms? *American Mathematical Monthly* 106, 99–111.

Feferman, S. (2004a) Tarski's conception of logic. *Annals of Pure and Applied Logic* 126, 5–13.

Feferman, S. (2004b) Tarski's conceptual analysis of semantical notions. *Sémantique et épistémologie* (A. Benmakhlouf, ed.). Editions Le Fennec, Casablanca [distrib. J. Vrin, Paris], 79–108.

Feferman, A. B. and Feferman, S. (2004) *Alfred Tarski. Life and Logic*. Cambridge University Press.

Fernández Moreno, L. (1990) Tarski and the concept of logical constant. *Logique and Analyse* 131–132, 203–214.

Ferreirós, J. M. (1999) *Labyrinth of Thought. A History of set theory and its role in modern Mathematics*. Birkhauser, Basel.

Field, H. (1972) Tarski's theory of truth. *Journal of Philosophy* 69, 347–375.

Frost-Arnold, G. (2004) Was Tarski's Theory of Truth Motivated by Physicalism? *History and Philosophy of Logic* 25, 265–280.

Givant, S. (1991) A portrait of Alfred Tarski. *Math. Intelligencer* 13, 16–32.

Givant, S. (1999) Unifying threads in Alfred Tarski's work. *Math. Intelligencer* 21, 47–58

Gödel, K. (1986) *Collected Works II*, eds. S. Feferman et al. Oxford University Press, Oxford.

Gödel, K. (1995) *Unpublished philosophical essays*, F. Rodríguez-Consuegra (ed.). Birkhäuser, Basel, Boston and Berlin.

Gómez-Torrente, M. (1996) Tarski on logical consequence. *Notre Dame Journal of Formal Logic* 37, 125–151.

Gómez-Torrente, M. (2001) Notas sobre el "Wahrheitsbegriff", I and II. *Análisis filosófico* XXI, 5–41 (I), 149–185 (II).

Haller, R. and Tarski, J. (1992) Alfred Tarski: drei Briefe an Otto Neurath. *Grazer Philosophische Studien* 43, 1–32.

Hintikka, J. (1988) On the development of the model-theoretic viewpoint in logical theory. *Synthese* 77, 1–36.

Hintikka, J. (2004) On Tarski's assumptions. *Synthese* 142, 353–369.

Jané, I. (1993) A Critical Appraisal of Second-Order Logic. *History and Philosophy of Logic* 14, 67–86.

Jané, I. (2001) Reflections on Skolem's Relativity of Set-Theoretical Concepts. *Philosophia Mathematica* 9, 31–55.

Jané, I. (2004) De qué trata la teoría de conjuntos?. In Orayen, R. and Moretti, A. (eds.) (2004), 247–276.

Kanamori, A. (1996) The mathematical development of set theory from Cantor to Cohen. *Bull. Symbolic Logic* 2, 1–71.

Kanamori, A. (1997) *The Higher Infinite: Large Cardinals in Set Theory from their Beginnings.* Springer-Verlag, Berlin and New York.

Kotarbinski, T. (1955/1935) The fundamental ideas of pansomatism. Translated into English by A. Tarski and D. Rynin. Reprinted in Tarski, A. (1986), vol. 3, 579–591.

Krajewski, S. (2003) Gödel and Tarski. *Annals of Pure and Applied Logic* 126, 303–323.

Kreisel, G. (1967) Informal rigour and completeness proofs. In Lakatos, I. (ed.) (1967), 138–171.

Lakatos, I. (ed.) (1967) *Problems in the Philosophy of Mathematics.* North- Holland, Amsterdam.

Lévy, A. (1988) Alfred Tarski's work in set theory. *Jrn. Symbolic Logic* 53, 2–6.

Maddy, P. (1996) Set theoretic naturalism. *Journal of Symbolic Logic* 61, 490–514.

Mancosu, P. (2005a) Tarski on models and logical consequence. In J. Gray and J. Ferreiros (eds.), *The Architecture of Modern Mathematics.* Oxford University Press, forthcoming.

Mancosu, P. (2005b) Harvard 1940–41: Tarski, Carnap and Quine on a finitistic language of mathematics for science. *History and Philosophy of Logic*, forthcoming.

Mostowski, A. (1967a) Recent results in set theory. In Lakatos, I. (ed.) (1967), 82–96.

Mostowski, A. (1967b) Tarski, Alfred. In *Encyclopedia of Philosophy*, P. Edwards ed.. Macmillan, New York, vol. 8, 77–81.

Mycielski, J. (2003) On the Tension Between Tarski's Nominalism and his Model Theory (Definitions for a Mathematical Model of Knowledge). *Annals of Pure and Applied Logic* 126, 215–224.

Niniluoto, I. (1999) Theories of truth: Vienna, Berlin, and Warsaw. In [72],1726.

Orayen, R. and Moretti, A. (eds.) (2004) Filosofía de la lógica. *Enciclopedia Iberoamericana de Filosofía*, vol. 27, Trotta, Madrid.

Rodríguez-Consuegra, F. (1989) Russell's theory of types, 1901–1910: its complex origins in the unpublished manuscripts. *History and Philosophy of Logic* 10, 131–164.

Rodríguez-Consuegra, F. (1991) *The mathematical philosophy of Bertrand Russell: origins and development*. Birkhäuser, Basel, Boston and Berlin (reprinted in 1993).

Rodríguez-Consuegra, F. (1995) Kurt Gödel and the philosophy of mathematics. Introductory essay in Gödel, K. (1995), 17–106.

Rodríguez-Consuegra, F. (2004a) Tipos lógicos, lenguaje y filosofía. In Orayen, R. and Moretti, A. (eds.) (2004), 217–228.

Rodríguez-Consuegra, F. (2004b) Propositional ontology and logical atomism. In G. Link (ed.), One Hundred Years of Russell's Paradox. W. de Gruyter, Berlin, 417–434. [Reprinted n *this* vol., ch. 10.]

Rojszczak, A. (2002) Philosophical Background and Philosophical Content of the Semantic Definition of Truth. *Erkenntnis* 56, 29–62.

Sagüillo, J. M. (1997) Logical Consequence revisited. *Bulletin of Symbolic Logic* 3, 216–241.

Simons, P. (1987) Bolzano, Tarski, and the Limits of Logic. *Philosophia Naturalis* 24(4), 378–405.

Sinaceur, H. (2000) Address at the Princeton University Bicentennial Conference on problems of mathematics, by A. Tarski. *Bulletin of Symbolic Logic* 6, 1–44.

Sinaceur, H. (2001) Alfred Tarski: Semantic Shift, heuristic shift in metamathematics. *Synthese* 126, 49–65.

Suppes, P. (1988) Philosophical implications of Tarski's work. *J. Symbolic Logic* 53, 80–91.

Tarski, A. (1938) Über unerreichbare Kardinalzahlen. *Fundamenta Mathematicae* 30, 68–89. Reprinted in Tarski, A. (1986), vol. 2, 359–380.

Tarski, A. (1939) On well-ordered subsets of any set. *Fundamenta Mathematicae* 32, 176–183. Reprinted in Tarski, A. (1986), vol. 2, 551–558.

Tarski, A. (1944) The semantic conception of truth and the foundations of semantics. *Philosophy and Phenomenological Research* 4, 341–376.

Tarski, A. (1958) Remarks on predicate logic with infinitely long expressions. *Colloquium Mathematicum* 6, 171–176. Reprinted in Tarski, A. (1986), vol. 4, 11–16.

Tarski, A. (1962) Some Problems and Results Relevant to the Foundations of Set Theory. In E. Nagel, P. Suppes and A. Tarski (eds.), *Logic, Methodology and Philosophy of Science.* Stanford University Press, 125–135.

Tarski, A. (1983) *Logic, Semantics, Metamathematics*, J. Corcoran ed. Oxford University Press, Oxford. Second edition. (First edition, 1956).

Tarski, A. (1986) *Collected Papers*, S. Givant and R. McKenzie eds., vols. I–IV. Birkhäuser, Basel.

Tarski, A. (1986/1966) What are logical notions? J. Corcoran ed. *History and Philosophy of Logic* 7, 143–154.

Tarski, A. (1987/1944) A philosophical letter of Alfred Tarski. A. White ed. *Journal of Philosophy* 84, 28–32.

Tarski, A. (1995/1939–40) Some current problems in metamathematics. J. Tarski and J. Wolenski eds. *History and Philosophy of Logic* 16, 159–168.

Tarski, J. (1996) Philosophy in the creativity of Alfred Tarski. *Dialogue and Universalism* 6, 157–159.

Wang, Hao (1996) Skolem and Gödel. *Nordic Journal of Philosophical Logic* 1, no. 2, 119–132.

Wolenski, J. (1993) Tarski as a philosopher. *Poznan Studies in the Philosophy of the Sciences and the Humanities* 28, 319–338.

Wolenski, J. and Köhler, E. (eds.) (1999) *Alfred Tarski and the Vienna Circle : Austro-Polish connections in logical empiricism*. Kluwer, Dordrecht.

Departamento de Lógica y Filosofía de la ciencia
Universidad de Valencia

Chapter 13
From Logic to Philosophy

Abstract I've been studying and researching the work of some of the greatest logicians of the XX century, including their philosophical implications, for many years. As usual in the professional publications, one must leave personal relationships aside, to concentrate in their achievements. Here I try to cover that more personal side, to explain, to some degree, how and why my interest in those logicians started and evolved. At the same time, I'll try to point out the way in which some important philosophers were inspired by logic to try to reach their goals. Thus, this last chapter, written especially for this compilatory volume, may be seen as a complement to my many publications on that field, and very especially to this book.

Keywords B. Russell · K. Gödel · A. Tarski · W.V. Quine · Biography · Logic · Philosophy

Mathematics Subject Classification Primary 01A70 · Secondary 03A05

Logicians are great people, that's what I thought as soon as I started to study analytical philosophy. Without a strong basis in logical studies, it's very difficult to get a precise language, and clear ideas and argumentation. What is very easy to get is literature, instead of useful philosophy. In the following I'll briefly describe the way which took me to some of those philosophers who really knew logic.

F. Rodríguez-Consuegra, *Russell, Gödel, Tarski*, Studies in Universal Logic,
https://doi.org/10.1007/978-3-032-08945-8_13

13.1 Bertrand Russell

The work of Bertrand Russell was the main reason why I decided to study Philosophy. I still remember my excitement in trying to understand, as a very early teen, his popular *The Problems of Philosophy*. I knew about some of his logical writings and was fascinated by what I could read about his great work *Principia Mathematica* (by then I even ignored the role of Whitehead as a coauthor). Yet after high school I decided to become a Merchant Marine officer, mostly due to my intense desire to travel and see the world. As soon as I graduated and started sailing professionally, I started my second degree: philosophy. It took me a few years, but in the meantime I had lots of time to read (almost devour) many books, mostly in the field of analytical philosophy. My project to meet Russell someday was frustrated when I knew, while visiting Buenos Aires in 1970, that he died.

Very soon I knew that my doctoral dissertation had to be devoted to Russell's philosophy. I chose a subject which allowed me to delve deep into the whole of Russell's work: his analytical methods. Then I discovered that Russell sold his archive to the Hamilton University, in Canada, so I contacted Ken Blackwell, the archivist. I got a catalogue of the materials available in the Archives, so I ordered lots of xerocs of the unpublished manuscripts which seemed to me important to try to reconstruct Russell's development respecting the subjects I was interested in, mostly between 1898 and 1910. To prepare and compose my doctoral dissertation took me about four years, and the result was a very thick thesis of 800 pages, which was finished and presented in 1988. As I said, it was an attempt to isolate the main traits of Russell's analytical methods along his whole philosophical development.

It was already time to start publishing some results of my work, which I did first in *History and Philosophy of Logic*, the prestigious academic journal founded by Ivor Grattan-Guinness, who supported my work, even when my name was still very little known. Our relationship lasted for many years, until he died in 2014. We met many times, in several countries, and he was a very good friend of mine as well.[1] Below there is a pic of us together in London, October 1989, when I was ready to travel to Canada, to spend two semesters in the Hamilton University, to study the Russell Archives:

[1] I can say something similar about John Corcoran. I devoted a whole paper to him: "My personal memories of John Corcoran". In Jean-Ives Beziau and Tim Madigan (eds.), *Universal Logic, Ethics, and Truth: Essays in Honor of John Corcoran (1937–2021),* Studies in Universal Logic, Springer, 2024, pp. 151–164.

The Russell Archives is a fantastic place to study and research. Even emotionally it is impressive as well, since you can enjoy Russell's private library, which was sold to McMaster University together with the rest of the unpublished materials, mostly manuscripts and correspondence. You can see also Russell's desk, where he used to write.

Then everything was faster, Birkhäuser accepted my first book for publication, which appeared in 1991,[2] and many other publications started to appear in several countries, both in Spanish and English. Then Russell's philosophy was one of my main subjects of investigation for many years. Even in 2019 a second book of mine about Russell appeared, which was started while I was in Canada in 1989–90, and finished in Harvard in 1993. The manuscript of that book was apparently lost, but it appeared suddenly in an unexpected way.[3]

Through my work on Russell, I could somehow verify my idea that any good, sound, useful philosophy should be founded on logical basis. Russell liked to refer to his philosophy as the result of the application of a scientific method in philosophy, which culminated in his well-known logical atomism. In the following, other samples of philosophies based on logical foundations will be briefly considered.

---

[2] *The mathematical philosophy of Bertrand Russell: origins and development.* Basel, Boston and Berlin, Birkhäuser, 1991 (reprinted 1993). A whole updated list of my publications can be seen here: https://www.uv.es/rodriguf/publications.html.

[3] *Relational Ontology and Analytic Philosophy. Bertrand Russell and Bradley's ghost.* More about this book here: https://www.amazon.com/dp/1709974710.

13.2 Kurt Gödel

My interest in Gödel started when I was in Canada in 1989–90, devoted to study the Russell Archives. There I was invited by the Russell Society to give a lecture about my work. One of the sections of my lecture was devoted to somehow defend Russell of the popular accusation that he did not understand Gödel's famous incompleteness results. After my lecture, Gregory H. Moore, the famous historian of set-theory, who was in the audience, talked to me and advised me to visit the Gödel *Nachlass*,[4] in Princeton University. This I did, which led me to open another line of research, which lasted for a few years, and resulted in quite a few papers and books, in Spanish and English.[5]

I spent in Princeton about a week in the Spring of 1990, after my stay in the Russell Archives, Canada. Curiously enough, I stayed in *The Peacock Inn*, the same place where Gödel and Russell spent some time in the past. Sadly, the owner was unable to give me details about those old stays. In my first visit to the Firestone Library, where the *Nachlass* was permanently located, I was totally astonished about the richness and variety of the Gödel materials contained there.

Since my goals were mostly philosophical, I concentrated in the unpublished manuscripts of a philosophical character. These were the Gibbs Lecture, given at Brown University in 1951, and then up to six versions of a paper entitled "Is Mathematics Syntax of Language?". These versions were written between 1953 and 1959, when Gödel was invited to participate in the Schilpp volume devoted to Carnap's philosophy (published in 1963), but none of them was finally submitted for publication. The most fascinating trait of that paper is that the different versions apparently suffered a process of self-censorship, since the first version has more than 100 folios and the last one about 6 folios. Most likely Gödel was not satisfied with his arguments against Carnap and did not want to be involved in public controversy. I got xerocs of the Gibbs lecture and of several of those six versions, and devoted several years to reconstruct them, then publish my results in several papers and books, as mentioned above.

In 1992 I met Hao Wang, the celebrated logician, in the Wittgenstein Symposium on Philosophy of Mathematics, Kirchberg, Austria. He was the only person who was authorized by Gödel to maintain deep philosophical conversations with him and take notes about them. We had a very long conversation about Gödel's philosophical ideas and interchanged our projects related to Gödel's philosophy. I was working in what would become my first book about Gödel's unpublished manuscripts, and he was working on the notes he took on those conversations with Gödel, so we really had a lot of information and ideas to share on that fascinating subject.[6]

---

[4] A full information about Gödel's archive here: https://books.openedition.org/pup/53545.

[5] My most important book in English was this: *Kurt Gödel, Unpublished philosophical essays*. Basel, Boston and Berlin, Birkhäuser, 1995. The original Spanish version appeared in 1994: Barcelona, Mondadori. As for the papers devoted to that subject, most of them are collected in this volume, Chaps. 3–7 and 9.

[6] See Chap. 9 of this book, "Philosophy in Hao Wang's conversations with Gödel".

My main thesis about those unpublished materials was, and still is, that Gödel was not only a logician, but also, and perhaps mainly, a philosopher looking for technical results which could be used to support his Platonistic philosophy of mathematics.[7] So Gödel can be clearly added to my short list of philosophers who based their philosophic positions and arguments on technical, apparently irrefutable logical or mathematical arguments or results.

13.3 Willard V. Quine

When I knew about Russell's passing away I was very young, just 19 years old. The first thing I thought was: unfortunately, I will never be able to meet him. That was a depressing thought for me, since Russell was my main inspiration to study philosophy. In a few years I started to greatly admire Quine's philosophy as well, even as much as I admired Russell's. Then I thought: perhaps I will be able to meet Quine! This time I could manage to make my dream come true.

The most important thing I learned about Quine was that he based his main philosophical doctrines in strong logical arguments. This is not easy to see, but once you understand it, almost all his most famous philosophical theses can be derived (in a rather weak sense) from what we can consider his starting point: "to be is to be the value of a variable". That is, indeterminacy of translation, ontological relativity, inscrutability of reference, and even the celebrated attack against the analytic-synthetic distinction, are all consequences of that starting point.[8]

A little before I finished my Ph.D. dissertation, in 1987, I sent a letter to Quine trying to consult him about some doubts of mine, with the hope that he perhaps could find some time to read my letter and even reply. In particular, I sent to him a long essay about Peano, Frege and Russell (in Spanish, since I knew Quine was a polyglot), asking for comments. You can easily imagine my excitement when a handwritten letter by Quine reached me in a matter of a few weeks. He wanted me to be more precise about what particular passages of my long paper I wanted him to study and comment. Needless to say, I did it immediately, which led Quine to send me a couple of more letters, this time typewritten, with a lot of details and arguments.

Our main subject of disagreement was the role of Peano in Russell's discovery of his celebrated theory of descriptions. Quine insisted that Peano limited his role to the use of a convenient notation, otherwise Peano believed that descriptions are irreducible, so they cannot be eliminated, as Russell did so efficiently. I managed to give Quine all the details, precisely showed that the same was done by Peano, with very clear logical expressions, even citing the most important passages of Peano's

[7] My main publications on this subject have been studied, and compared among them, by Fernando Zalamea in his paper "Kurt Gödel: análisis filosófico y lógica matemática". *Mathesis* 12/1996, pp. 347–374. This paper can be downloaded here: https://www.uv.es/rodriguf/Zalamea-Godel.pdf

[8] I did it in my "La reducción ontológica y sus problemas". *Crítica* 70 (1992): 17–64. The manuscript of this paper was also read by Quine, who sent me useful comments.

published writings. The next letter by Quine contained this spectacular passage, which delighted me for a long time: "I am happy to get straight on Peano on descriptions. I checked your reference, and I fully agree. Peano deserves all the credit for it that has been heaped on Russell (except perhaps for Russell's elaboration of the philosophical lesson of contextual definition)." I was very happy to cite this passage in print when I devoted a full paper to that subject.[9]

Our correspondence lasted for a few years, until 1993, when I got a Fullbright Fellowship to spend two semesters in Harvard. (As a matter of fact, Quine himself encouraged me to go there.) Yet I was so fortunate to meet Quine before. In 1990 he was invited to lecture in Girona (Spain), so I was there to listen to him and meet him personally. I was astonished when, as soon as the organizers saw me, they told me: "Very little after his arrival here, Quine was asking about you!". So, I had the great pleasure to shake hands with him and talk for a while, since I was unable to attend all his lectures, due to my work (I was teaching philosophy in high school).

In 1992 Quine kindly accepted to write a Foreword for my edition of some of Gödel's unpublished manuscripts of a philosophical character. It appeared both in the Spanish and English editions.[10] He was very interested in Gödel's ideas and was very generous with his time, so I could not be more grateful.

Once in Harvard, I was so fortunate to talk with Quine several times. He was already retired, but he liked to visit his office in the Philosophy Department from time to time, especially with good weather. My impressions in being seated in front of him was unforgettable: I saw him as a fabulous giant, probably the most important philosopher of the XX century. The first day he immediately showed me the thick file where he kept my letters and papers. Yet I was mostly interested in talking about personal matters of his productive life, like his relationships with Carnap and Russell, etc. However, he used to give me kind but short answers to this sort of questions, then to insist: "Let's talk philosophy". He was 85 years old, and certainly looked like that, but he still wanted to talk philosophy!

Two small details surprised me. At the entrance in his office, I saw a framed photo of Frege hanging on the wall. He confirmed me his admiration for the great logician. The other detail was about his way to write his papers, which he showed me with some pride. He took notes in different pieces of paper. Then he used scissors to cut them and placing them in the right order by gluing them to clean sheets. Quine was very prolific, so we can only speculate about his output if he had been able to use a computer.

Quine died in 2000, then I wanted to do something to increase the knowledge of his philosophy in Spain, as a little homage of mine to his memory. So, I chose a few papers by him not previously translated into Spanish, translated them and publish a book in 2001, with a quite complete introduction to his philosophy. I sent a copy of that book to Quine's son, Douglas, who maintains a big website about his father's

[9] "Frege, Peano and Russell on descriptions: a comparison". *Russell. The Journal of Bertrand Russell Studies* 20 (2000): 5–25.

[10] See footnotes 5 and 7 for details.

life and philosophy, where all the publications about Quine that continue to appear non-stop are duly registered.[11]

13.4 Hilary Putnam

If there ever existed a logical argument from where a philosophical conclusion could be “derived”, that was Putnam’s model-theoretic argument against metaphysic realism, especially as it appeared in his “Models and reality”.[12] The presentation of the argument by using a formalized language is complex, but the gist of the argument comes to this: no criterion which determine the truth values of the statements of a language is also able to determine the reference of its terms. In other words, our statements are semantically indeterminate (there is no “facts of the matter” about them). Thus, any correspondence theory of truth is hopeless, together with any form of strong (metaphysical) realism. That situation led Putnam to his celebrated “internal realism”.

The argument has been analyzed and discussed for decades, even after Putnam renounced to internal realism, but for me it was the reason why I contacted Putnam. As soon as I discovered that the famous paper was not yet available in Spanish, I decided to organize the translation, with the collaboration of Roberto Torretti, and starting from a former unpublished translation by a colleague who preferred to remain anonymous. Putnam gave us full permission, and he even wanted to take advantage to write a short complementary paper to somehow clarify some aspects of the original publication.[13]

In August 1992 I was invited to participate in the big Putnam Symposium, Taxco, Mexico, organized by the IIF-UNAM. That was a beautiful event, in a fantastic scenery, plenty of great philosophers. There I met Putnam personally for the first time, so I had the opportunity to discuss with him the details of the mentioned Spanish translation, about which he was delighted.

However, Putnam was rather unhappy with that event, since almost all the presentations were critical with different aspects of his philosophy. Some of the criticisms involved a sort of accusation about the incoherences which could be found when you compare different publications by Putnam, which led the famous philosopher to declare: “I am a piecemeal philosopher”. That sounded as a sort of excuse, but in any case it left us rather stunned.

---

[11] My book: *W. V. Quine, Acerca del conocimiento científico y otros dogmas.* Barcelona, Paidós, 2001. The Quine site is here: https://www.wvquine.org/.

[12] *Journal of Symb. Log.*, no. 45 (1980), pp. 464–482.

[13] The translation appeared in *Diálogos* 63 (1994) pp. 7–45, followed by Putnam’s “Afterthoughts on ’Models and Reality’”, in English and Spanish.

I gave a presentation highly critical as well.[14] By comparing Putnam with Quine about the inscrutability of reference, and related subjects, I tried to show that Putnam's criticisms of Quine were unconvincing and unfounded, and even incoherent. Also, I gave arguments to show that Putnam's position didn't go much further than Quine's.

Putnam was very close to me, seated at my right hand, in the same very long table, and while I was reading my text I noticed that he didn't like my arguments at all, so I was prepared for a hot discussion. To make a long story short, Putnam criticized my position with a variety of arguments, to which I replied with new attacks against him. Finally, we closed the discussion, and the old fox looked at me and said: "You didn't catch me", with a big smile in his face. As a sort of compliment, I replied "Yep, you're really clever", which was followed by a great applause from the audience.

A very funny anecdote took place when the local Taxco newspaper wanted to publish a photo of all the participants in the Putnam event. Unfortunately, by the same dates a beauty miss contest was taking place too. When the printer wanted to inform about both events in the front page, someone inserted the wrong captions for each photo. The results could not be more exhilarating. This is one side of the confusion:

El Instituto de Investigaciones Filosóficas de la Universidad Nacional Autónoma de México celebró en el Hotel Monte Taxco su XI Simposio de Filosofía: En Torno a la Obra de Hilary Putnam. La presente gráfica muestra a todo el grupo que asistió. (Foto Arroyo)

[14] I gave my presentation in English, but since Putnam did not want the projected book containing all the presentations, with his replies, to be published, mine was published in Spanish: "Putnam, Quine y la 'facticidad' de la semántica". *Diánoia* (1992): 171–185.

As I said above, in 1993 I spent time in Harvard, and I took advantage to briefly talk to Putnam in a couple of occasions, one of them related to Wittgenstein. He showed himself as a good connoisseur of the Austrian genius. Also, I let him know that the promised translation of his "Models and reality" was already close to appear. He was very kind to me. As it happened, the only Mac computer available in the Philosophy Department was his, and I was authorized to use it, which allowed me to finish a book I had been working on for a long time, and print a full copy.[15] Also, Putnam invited me to have lunch together in the Harvard Faculty Club, where I took advantage to make quite a few comments about his last publications, mostly to show him that I was still interested in his work, which, by the way, was not totally true.

13.5 Alfred Tarski

My research interest in Tarski came from several sources. First, through my discovery that Tarski's famous definition of logical consequence was anticipated by members of the Peano school.[16] Second, through my study of the relationship between Gödel's results and Russell's philosophy, very especially his logicist foundations of mathematics. This led me to delve deep in what we know as Tarski's theorem: that arithmetical truth cannot be defined in arithmetic. After some research, I came to belief that the celebrated theorem was first formulated by Gödel, in an especially convoluted way, which led me to write a full paper on this subject (Chap. 11 of this volume). All in all, I decided to visit the Tarski Papers, in Berkeley (that was my third way to Tarski), where I discovered interesting materials of philosophical interest.[17] Later, I devoted much time to try to understand Tarski's philosophical views on set theory, which resulted in writing a long a paper: "Tarski's intuitive notion of set", published in 2005, and reprinted in this volume (as Chap. 12).

My stay in Berkeley as a Fullbright Visiting Scholar (Mathematics Department, June 1996) was possible thanks to the financial support of a Spanish Research Project and to Leon Henkin, who officially invited me to go there, made all the arrangements needed and assisted me in many ways. Also, I could enjoy the personal help of Charles Chihara, John Addison and Jan Tarski, as well as of Solomon Feferman and Patrick Suppes, in correspondence.

Henkin introduced me to the Maths Dept, where they were so kind as to provide me with an office and full access to internet through the computers room. Also, Henkin wanted to show me his office, which he no longer used after his retirement.

---

[15] See footnote 3 for details about that book.

[16] Full details about that, in the context of the Peano School, can be found in the third chapter of the book, coauthored with E. Marchisotto and J. Smith, *The Legacy of Mario Pieri in Foundations and Philosophyof Mathematics*. New York and Basel, Birkhäuser, 2021. The contents of this book can be seen here: https://link.springer.com/book/10.1007/978-0-8176-4823-7.

[17] Some of those materials we published here: "Two unpublished contributions by Alfred Tarski." *History and Philosophy of Logic* 28 (2007), pp. 257–264.

The desk was plenty by a big mountain of letters and other documents. With a little rogue sunrise he said that now he didn't have time to devote to such things.

On a more personal side, Henkin invited me to dinner at his home. The house was located in a spectacular place, from where you could enjoy a fantastic scenery, especially because some of the walls were made of glass. He showed me his home office, where I saw another big mountain of letters and other documents on his desk, which clearly were awaiting attention. In the top of the mountain, I saw a copy of my book on Russell mathematical philosophy,[18] which doubtless was located there as a nice gesture to me. He and Ginette, his kind wife, treated me all the time as if we were old friends, so as a whole the event was something really unforgettable to me.

Yet Henkin did something else for me a few years later, when he was so kind and generous as to write and send a letter to my university, supporting my candidacy for a full professor position. I felt sad when I knew that he and his wife left the house and moved to Oakland to live in a nursing home, and even more when Henkin died in 2006. Unfortunately, I was unable to send him a copy of my paper about Tarski's philosophy.

Finally, I had the opportunity to meet Tarski's son, Jan Tarski. He invited me to dinner in a nice restaurant, close to the ocean, where he was so kind as to answer many questions about his father and his family. Also, he took me to see the house where the Tarski family lived for a long time.

The Bancroft Library, where the Tarski Papers were located, gave me all the support needed so that I was able to start researching all the materials extant there. Box after box, file after file, I needed no less than three weeks to check and revise all the documents available. I was not looking for anything in particular. I just wanted to see if I would be able to try to confirm my suspicion that some documents with a philosophical interest might be found there. As I said above I was right about my suspicion.

The most important document from the philosophical viewpoint was the text of an unpublished contribution to a Chicago meeting in 1965. A single quotation of that contribution may constitute a summary of his philosophical position:

> I represent this very rude kind of anti-Platonism, one thing which I could describe as materialism, or nominalism with some materialistic taint, and it is very difficult for a man to live his whole life with this philosophical attitude, especially if he is a mathematician, especially if for some reasons he has a hobby which is called set theory, and worse – very difficult. (Tarski, Chicago, 1965)

And that contribution was the one I chose to be read when Henkin organized a meeting with himself, Charles Chihara, the philosopher, and John Addison, the set-theorist. We were seated around a table, then Henkin told me something which impressed me a lot: they had me seated in the same chair used by Tarski in their meetings!

[18] See footnote 2.

The goal of the meeting was to give me the opportunity to tell my experience with the Tarski papers. They were rather astonished to hear the details of Tarski's philosophical position, expressed in so clear terms. Then we were discussing some of the arguments, and all of them made very useful comments about them. I took careful notes, which I was able to use when I wrote the long paper mentioned above (reprinted in this volume, as Chap. 12).[19]

[19] When Hilary Putnam knew about the existence of the text of Tarski's Chicago 1965 contribution, he asked me to send a copy to him. Once he read the text, he declined to make comments about it, despite my suggestion. His only reason was: "Tarski was no philosopher".

GPSR Compliance
The European Union's (EU) General Product Safety Regulation (GPSR) is a set of rules that requires consumer products to be safe and our obligations to ensure this.

If you have any concerns about our products, you can contact us on

ProductSafety@springernature.com

In case Publisher is established outside the EU, the EU authorized representative is:

Springer Nature Customer Service Center GmbH
Europaplatz 3
69115 Heidelberg, Germany

www.ingramcontent.com/pod-product-compliance
Lightning Source LLC
Chambersburg PA
CBHW071447171125
35558CB00002B/19
* 9 7 8 3 0 3 2 0 8 9 4 4 1 *